COMPUTERIZED
FOOD PROCESSING
OPERATIONS

COMPUTERIZED FOOD PROCESSING OPERATIONS

Arthur A. Teixeira
Associate Professor of Food Engineering
University of Florida

Charles F. Shoemaker
Associate Professor of Food Science
University of California, Davis

Springer Science+Business Media, LLC

An AVI Book
(AVI is an imprint of Van Nostrand Reinhold)
Copyright © 1989 by Springer Science+Business Media New York
Originally published by Van Nostrand Reinhold in 1989

Library of Congress Catalog Card Number 88-17415

ISBN 978-0-442-28501-2   ISBN 978-1-4615-2043-6 (eBook)
DOI 10.1007/978-1-4615-2043-6

Library of Congress Cataloging-in-Publication Data

Teixeira, Arthur A., 1944–
    Computerized food processing operations.

    "An AVI book."
    Bibliography: p.
    Includes index.
    1. Food industry and trade—Data processing.
I. Shoemaker, Charles F., 1946–   II. Title.
TP370.5.T45   1989   664'.02'0285   88-17415
ISBN 978-0-442-28501-2

To Our Wonderful Families

Marjorie
Alan, Craig, Scott

Sharon
John, Robert

CONTENTS

PREFACE

This book is designed to explain and illustrate how food processing operations can be made more efficient and profitable through the application of computers in the laboratory, pilot plant, and production plant floor of industrial food processing plants. It is intended to provide a sufficient understanding of how computer system concepts can be applied to food processing operations to permit technical managers, with the assistance of food engineering professionals, to identify, develop, and implement computer applications to meet their own specific needs. The book should also serve as a useful text or guide for students in food engineering or food technology seeking a practical course on food process automation at the undergraduate-graduate level interface.

The material covered includes the use of microcomputers for automated data acquisition and analysis in the laboratory and pilot plant, followed by the use of computer-based process control systems on the production plant floor. Higher-level applications are also included to illustrate the use of engineering software containing mathematical models for computer simulation, optimization, and intelligent on-line control of specific food processing unit operations.

In each chapter, an introduction to the theory of the application in simple lay terms is followed by case study examples of actual project installations or demonstration projects that illustrate the application in a specific food processing situation. Some of the unit operations covered include thermal processing in retorts, heat treatments in aseptic processing, freezing, dehydration, evaporation, and fermentation. Product examples span across food industry subsectors from dairy products to distilled spirits.

ACKNOWLEDGMENTS

Thanks are due to many individuals for supplying information and illustrations. Particular thanks go to Dr. Dennis R. Heldman of the National Food Processors Association for encouraging the work and for his review of the manuscript. The authors are also grateful to Dr. John E. Manson of Central Analytical Laboratories, Inc., for his many suggestions and comprehensive review of chapter sections dealing with thermal process sterilization. Special thanks also go to Dr. William C. Dries of the University of Wisconsin's Department of Engineering Professional Development for including the principal author as a regular speaker in his annual short-course on "Computer Control of Food Processing," which provided much of the material in Chapter 3. For the remaining material in the book the authors are grateful for their combined industrial and academic experience in the field of food processing and food engineering that has made this work possible.

Introduction

A review of world history over the past thousand years will reveal certain periods of revolutionary changes that have had a major impact on modern civilization. The Renaissance brought Western Europe out of the Dark Ages into a quest for new knowledge. This quest led to exploration of the globe and to the establishment of a mercantile trade that brought great wealth to Europe along with the colonization of the New World. The Industrial Revolution of the nineteenth century transformed the United States from a rural agrarian economy to the industrial giant of the free world that it is today. The U.S. food processing industry has likewise experienced tremendous growth over the past hundred years as a result of the American strength in manufacturing technology that grew from the Industrial Revolution.

As the twenty-first century approaches, U.S. manufacturing industries, including food processing, are discovering that their share of both domestic and world markets is shrinking in the face of growing competition from Japan and Western Europe. The key to this growing foreign competition seems to lie in the use of sophisticated information systems that focus on quality assurance while achieving manufacturing efficiencies that produce better-quality products at lower cost. Far-thinking industrial analysts are pointing to these developments as evidence of a revolutionary change in our modern civilization that is being labeled the "information revolution." It is no surprise that what lies at the heart of this new information revolution are the high-speed digital computer and the dramatic developments in electronics technology that have encapsulated enormous quantities of computing power onto tiny silicon chips.

The potential impact that this revolution can have on American industry, and on food processing in particular, is limited only by the imagination of those who understand not only what computers can do but also the science and technology behind their industrial manufacturing operations. One purpose of this book is to help spark that imagination.

As computers became increasingly available, the food processing industry along with other major industries quickly adopted their use in the many business applications that were common across industry sectors. These included record keeping, accounting, payroll, least-cost ingredient formulations, database management, and management information systems. Although these applications were helpful in improving business management

1

efficiency, they had no direct impact on product and process development in the industrial laboratory or on industrial processing and manufacturing operations on the production plant floor. These are highly technical applications that need to be custom-developed by teams of engineers and scientists whose combined knowledge includes product and process technology, engineering mathematics, and computer science. Software for these applications can rarely be purchased as off-the-shelf packages from a vendor.

Dramatic improvements in laboratory efficiency can be achieved in the food industry laboratory by collecting primary data from sensors and instruments through a computer-based data acquisition system. These systems are capable of collecting, treating, and analyzing the data to produce results in a form ready for decision making by the research scientist. Computer-based process control systems can replace traditional hard-wire relay-logic control systems on the production plant floor, where they permit process modifications (such as changes in the sequence of unit operations) or addition or deletion of operations to be accomplished by simply reprogramming the computer, without the need for plant shutdowns to rewire the entire control system and process equipment interface.

Away from the plant floor, process design engineers can use sophisticated computer simulation software to predict the results of various process conditions on the end product. With this capability the engineer can make use of optimization techniques to determine the optimum set of process conditions that should be specified. Back on the plant floor, these same software packages can be accessed through the computer-based process control system. When unexpected changes occur in the course of a process, the simulation software can predict the outcome of the process as a result of these altered conditions. It can then calculate and implement downstream changes in process conditions that will compensate for the upstream deviation. In this way no off-specification product is produced. This is known as *real time on-line computer control*.

The foregoing examples illustrate the types of computer applications described in this book. It should go without saying that very specialized training is required for development of the sophisticated software involved. These computer programs are typically designed by food engineers (engineers with added training in food science) who have advanced training in engineering mathematics, engineering fields such as fluid mechanics and heat and mass transfer, and the reaction kinetics that describe the various physical, chemical, biochemical, and microbiological reactions occurring in food product systems. Much of this training lies beyond the reach of both the food scientist and the production or process design engineer who has only basic engineering training; yet these persons are normally responsible for recommendations leading to process improvement and automation for production efficiency in a food company. This book is intended to help such persons identify what can be done and the types of resources that would be necessary to do it. The authors hope that it will serve as a useful text for a college course in advanced food process engineering, as well as a continuing education text for practicing scientists and engineers in the food industry.

This book is about computerized food processing operations in the laboratory, pilot plant, and production plant floor. It provides a sufficient understanding of how computer system concepts can be applied to food processing operations to enable food science and engineering professionals (with the assistance of computer software engineers and systems specialists) to identify, develop, and implement computer applications to meet their own specific needs.

To meet this objective the material has been organized into six chapters. Chapter 1 introduces the reader to basic computer technology in common lay terms sufficient to provide an appreciation for what computer systems can do and why they can do it. In Chapter 1 special emphasis is placed on how computers can be interfaced with conventional laboratory instruments, sensors, and processing equipment so that the reader may appreciate the material covered in subsequent chapters. This chapter may be omitted by readers already proficient in computer technology.

Chapter 2 introduces computer applications in the laboratory for automated data acquisition in the chemical, physical, and microbiological quality analysis of food products and ingredients. Two sample case studies are given, involving gas chromatography for chemical analysis of food ingredients and heat penetration tests for designing thermal processes for canned food sterilization.

Chapter 3 takes the computer to the food processing plant for an introduction to the use of computer-based process control systems on the production plant floor. Case studies include dairy processing, vegetable oil refining, distilled liquor production, tank farm storage of citrus juice, and sterilization of canned foods.

On-line computer control of specific unit operations in food processing is presented in Chapter 4, with case studies of thermal processing, ultra-high-temperature sterilization-pasteurization, evaporation, fermentation, and drying. Each case study illustrates the importance of mathematical models for computer simulation of each unit process operation, which enables on-line decision making to adjust process conditions while processing is under way in real time.

Chapters 5 and 6 show how these process models have been developed and used to find processing conditions that maximize desired benefits while meeting necessary constraints. Case studies involving thermal processing, freezing, and drying are provided in these chapters.

The chapter sequence and content have been designed in such a way that the book should be useful for both the food scientist and the food engineer. For example, a course on computerized food processing designed for food science students would focus on Chapters 1 through 4. These chapters introduce the food science student to computer interfacing with laboratory and process equipment, computer-based process control systems, and on-line computer control of unit operations in food processing; a strong engineering background is not required in order to appreciate the concepts that are presented. On the other hand, a course designed for food engineering students would focus on Chapters 3 through 6, because Chapters 5 and 6 show how engineering science and mathematics are used to develop the computer programs for process simulation and optimization.

Chapter 4 thus serves as a fulcrum chapter. It describes the benefits that can accrue from on-line computer control of processing operations without going into detail on how the process simulation models are developed—Chapter 5 is reserved for that purpose. Chapter 4 awakens the reader to the powerful use that can be made of process simulation capabilities and should give the food scientist a new appreciation for the role that highly trained food engineers can play in the food industry. This chapter should serve to whet the appetite of the advanced food engineering student for mastering skills in process simulation and optimization that are covered in Chapters 5 and 6. The book could be used in a course designed for both food science and food engineering students by adjusting the depth of study assignments in each chapter accordingly.

An additional feature of the book is that one specific food processing operation (thermal sterilization of canned foods) has been used as a common case study example in each chapter. By following this example through each chapter, the continuity of successive computer applications dealing with the same operation will evolve. The reader will see how the increasing use of computer applications ranging from the laboratory to the production plant, along with behind-the-scenes software development, can lead to significant improvements in quality assurance and production efficiency for a specific processing operation.

Finally, the book's emphasis is on basic concepts and principles of application that should remain largely independent of continuing developments in computer hardware technology. Such advances may very well open the door to new applications not yet identified, however, and this book should serve as a stimulus in helping the food science professional to recognize such future opportunities. Although no attempt is made to identify all possible existing applications in food processing, case studies have been chosen from a sufficiently broad area that readers may determine how to approach other applications in order to meet their specific needs.

1

Microcomputers

Computers contain a series of miniaturized electronic circuits, which operate by storing and transmitting electrical signals. They are capable of accepting, storing, retrieving, and manipulating information. This ability to manipulate information in many different ways and store and retrieve it quickly makes computers valuable.

Before the computer can do its job, it needs to be loaded with a program. Programs, also called *software,* are very detailed instructions that tell the computer what to do. The computer and the software together constitute one unit, which is capable of doing a particular job. Some computers, such as the ones in aircraft autopilots, are dedicated to a single job; these computers always use the same program. Other computers can be used for many different tasks by simply changing their programs.

The computer is made of several major parts, which work together to manipulate and move information. It has a keyboard, very similar to a typewriter's, which allows entry of information; a video screen, which can display either letters and numbers in the form of a typed page or graphic images and pictures; and disk drives to "play" magnetic disks that store information in a form the computer can find quickly. Some personal computers use cassette tapes as their storage medium. Tape storage is slow and cumbersome for daily work but may be acceptable for some situations.

The computer has a central processing unit (CPU) and its associated integrated circuitry to do the "computing," which may entail calculating a value, searching for information on a disk, or just moving information from the keyboard into main memory. The CPU stores information electrically in circuits called *main* (or primary) *memory.* Main memory holds the program that tells the computer what to do and the data that the computer is using. The processor has rapid access to the information in the main memory. Additional information is stored on the disks, which are sometimes called *auxiliary memory.* It takes the processor more time to get information from a disk than from main memory, but the disk's storage capacity is greater. On some computers the processor, main memory, and auxiliary memory are housed together in a separate system unit.

Computers come in all sizes, from small microcomputers to large mainframe computers. The latter, which are primarily used for general-purpose data processing in business and scientific environments, were the first to be developed and were followed during the 1960s by *minicomputers,* which were smaller computers used in laboratories for data acquisition and general automation functions. They had much smaller memory sizes and were much slower at performing numerical calculations than mainframes. Later, in the 1970s, the microprocessor was developed, consisting of the entire CPU of a computer printed into one integrated circuit on a single silicon chip. Microcomputers built with these microprocessors became quickly competitive with minicomputers in computing power and have become the mainstay in most present-day laboratory and process automation applications.

Along with the development of compact and inexpensive microprocessors, new solid-state computer memory circuits were developed. These were much less expensive than the memories used in the memories of larger minicomputers or mainframes, which were made from magnetic core elements. In order to make this computing power easily accessible to users, various levels of software were developed, including programming languages, applications packages, and operating systems for microcomputers. Programming languages used for larger computer systems are also available for these smaller computers. New languages have even been developed specifically for microcomputers to take into account the limitations of microcomputer hardware and its integration into scientific and process control instrumentation. Along with the advances in hardware technology, software application programs such as word processing and electronic spread sheets have accounted for the widespread and rapid movement of microcomputer technology into our personal lives.

HARDWARE

Different microcomputer systems are in use in many types of applications. These systems are made with a variety of microprocessors, memory sizes, storage devices, and interface units; however, they contain common elements, which have similar functional relationships. A diagram of the most common elements of a microcomputer is presented in figure 1.1. These elements include a CPU, a memory, a mass data storage device, and input/output (I/O) devices. Intercommunication among these elements is accomplished over a set of parallel conductive paths called a *bus*. Expandable microcomputers provide for future additions to or extensions of the elements on a bus by including more memory, printers, and larger disk sizes, among other features. It is helpful to understand this structure when planning a project in which the computer is a major system component.

Central Processing Unit

The CPU is the main component of a computer. A microprocessor contains an entire CPU as an integrated circuit imprinted on a single *chip* (a chip is a small rectangular piece of silicon). A microprocessor CPU consists of

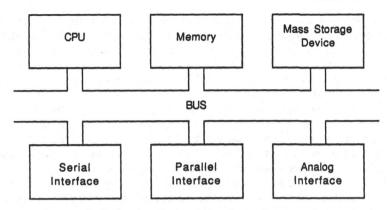

FIG. 1.1. A functional diagram of a microcomputer, showing the basic units: microprocessor or central processor unit, memory, mass storage device(s), interface ports, and the bus that provides a communication pathway among the units.

a variable number of registers and two units, the arithmetic logical unit (ALU) and the control unit. The specific design of a CPU varies among different manufacturers, but there are many common parts, and the overall modes of operation of early models were similar.

The ALU performs the arithmetic operations (such as addition and subtraction) and typical logic functions (including logical AND, logical OR, and shift operations). Closely associated with the ALU are registers called *accumulators* and a *status register*. An accumulator is a data storage register that holds operands before they are used in an ALU operation and also holds the results of operations after they have been performed. In addition to its computational ability, the ALU (when combined with a status register or *condition code register*) provides the microprocessor with the ability to make decisions based on the condition codes of the last ALU operation. A status register is a set of *flag* bits, which are used to represent operational result conditions. When a flag is set, the flag bit is at logic 1; when a flag is reset or cleared, the flag is at logic 0. The number or kinds of conditions that are associated with flag bits vary among different microprocessors. Conditions that normally have flag bits in microprocessors include arithmetic carries, arithmetic overflows, negative numbers, and zero.

For example, the addition of two numbers can produce a negative, zero, or positive result, for which the zero and negative flags would be 0,1; 1,0; and 0,0, respectively. Because of the fixed size of registers and accumulators, arithmetic and logic operations can also cause carries and overflows, which should be detected and corrected for proper results.

The second unit of a microprocessor CPU is the control unit, the role of which is to carry out, or instruct other parts of the microcomputer to carry out, the operational steps in the proper order. The proper operational steps are defined as a *program,* that is, a set of instructions that resides in the memory portion of the microcomputer. During the execution of a program, the control unit fetches, decodes, and executes each instruction in the proper order. In the fetch step, it fetches, or retrieves, the next instruction from memory and transfers it to the CPU. A special register, called the *program counter,* is used by the control unit to find the next instruction. The control unit is also responsible for updating the program counter before the next fetch step occurs. After fetching an instruction, the control unit decodes or interprets the desired action of the program instruction, and finally it executes the instruction by manipulating the parts of the microcomputer system that are necessary to perform the desired action.

If an arithmetic addition is programmed, then the ALU will be directed by the control unit to perform the operation. The execution of most instructions involve either the manipulation, movement, or testing of data. It is also the function of the control unit to carry out any movement of data, such as retrieval of a number from a particular memory location to be added to another already within the ALU. Along with the control unit and ALU of a microprocessor, there are a number of registers. Some are special-purpose registers such as the *program counter,* which always contains the address of the next memory location to be accessed by the control unit. Such registers have dedicated functions. Other registers in the CPU are general-purpose registers with several possible uses. A particular use at any given time is determined by the program. Uses of general registers include temporary data storage, addressing memory locations, and serving as accumulators with the ALU during computations.

One of the most widely used microprocessor chips is the 6502 microprocessor shown in figure 1.2. This CPU has one accumulator and a status register; it also has a program counter and two index registers. Certain of the 6502's instructions use the contents of an index register to extend the range of the address pointed to by the program counter. This CPU also has a stack pointer register, which is normally used for support of subroutine calls. For example, when a subroutine is called, the return address in the main program is saved at a location in memory pointed to by the contents of the *stack register.* The stack register also supports the *interrupt* structure of a CPU. This structural element allows an external device, such as an instrument, to request immediate service from the CPU. When interrupted, the CPU stops the execution of the current program and jumps or vectors to another program contained elsewhere in its memory to service the interrupt request. After completion of the interrupt program, the CPU will resume executing the program put aside by the interrupt request. The stack register is used by the CPU to find the next instruction to resume the execution of the main program.

The width, or number of bits, of the register of a microcomputer is important in evaluating its potential. The size of the program counter and that of the address bus interconnecting the CPU and other units determine how much memory can be addressed. The 6502 has a 16-bit program counter and address bus, which limits its address range to 2^{16}, or 64K (1K = 1024).

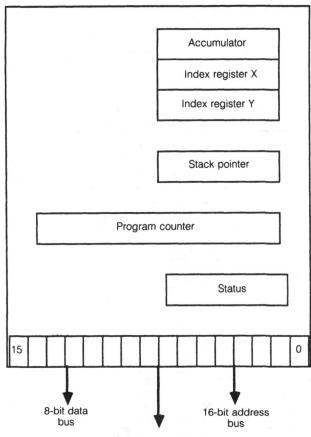

FIG. 1.2. MOS Technology's 6502 microprocessor. This microprocessor supports an 8-bit data bus and a 16-bit address bus. It has a limited number of registers, which are 8 bits wide except for the program counter, which is 16 bits wide . The 6502 was used in several of the early PCs, such as the Apple II and the Commodore 64.

The accumulator and data bus are 8 bits wide, which gives the 6502 its classification as an 8-bit microcomputer. More advanced CPUs have more and wider registers and also wider buses. For example, Motorola's 68020 microprocessor shown in figure 1.3 has eight data registers and seven address/index registers, which are all 32 bits wide. With a 32-bit address bus this CPU has an address range of 2^{32}, or 4 G-bytes. Wider buses and registers also provide for faster execution of programs. New types of registers speed up program execution and provide new functionality. For example, the 68020 has a kernel area designed for increased efficiency in multiuser and multitasking operating systems such as UNIX (Dessy, 1986b).

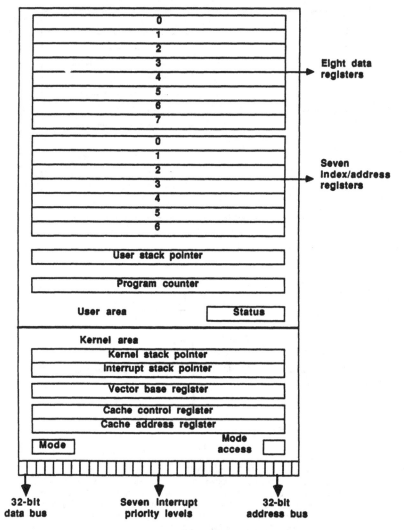

FIG. 1.3.   Motorola's 68020 microprocessor, which supports both a 32-bit data bus and an address bus. It has many different types of registers, which provides for increased efficiency and advanced functionality (Dessy, 1986b) over earlier 8- and 16-bit microprocessors.

Future architectural development of microprocessors is likely to emulate the architecture of the supercomputers of today. The CRAY X-MP/4 supercomputer (fig. 1.4) consists of four CPUs accessing the same memory, which has a 64-bit word size, and I/O devices. The multiple-processor configurations allow users to employ multiprogramming, multiprocessing, and multitasking techniques. The multiple-processor architecture can be used

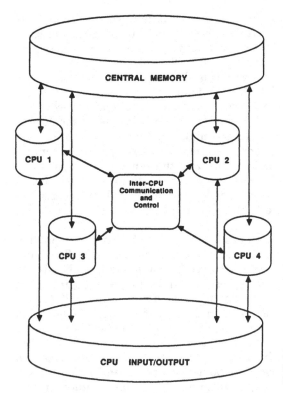

FIG. 1.4. The Cray X-MP/4 supercomputer has four CPUs and a 64-bit word size.

to process many different jobs for greater system throughput or to apply two or more processors to a single job for better program turnaround time. Multiprocessing and new techniques such as vector processing combine to provide a geometric increase in computational performance over conventional processing techniques.

Memory

The next unit of importance in a microcomputer is its memory, which provides storage for programs and data. When a program is being executed, the CPU fetches an instruction from memory and executes it. If the instruction being executed calls for the manipulation of data (also stored in the memory), then the CPU will fetch and manipulate that data from the memory. Computer memory may be thought of as a contiguous array of storage cells. Each cell has two attributes, its position in the array and its contents. The position of a memory cell is commonly called its *address*. Addresses make it possible for the CPU to distinguish each cell from the others. The standard size of an addressable cell for microcomputers has become the *byte*, that is, an ordered set of 8 binary digits (bits). A bit has only two possible states or values, 1 and 0, sometimes referred to as "on" and "off" or "true" and "false." A byte of memory will contain an ordered pattern of eight 1s and 0s. This pattern may represent an instruction or a piece of data. If it is an instruction, the CPU will fetch and execute it. Thus all possible operations that a CPU is able to perform are represented by a unique pattern of 1s and 0s. Many microprocessors require more than 1 byte for the representation of all their possible operations. For example, an instruction could have the CPU add one number located in memory to another number already in an ALU accumulator. Such an instruction may require 2 or more bytes for its representation. The first byte would contain the appropriate code to cause addition, and the remaining bytes would represent the address of the location in memory containing the number to be added to the accumulator. For example, the 6502, which has an address bus of 16 bits, would require 2 bytes for the representation of a 16-bit address.

Memory locations are also used to store data. Data that are manipulated by computer programs usually represent numbers or characters. The number of unique patterns that can be achieved by using 8 ordered bits is (2^8) or 256. Thus, if 1 byte is used to represent positive integers (beginning at 0 rather than 1), then the range 0 to 255 is possible. However, if 2 bytes (16 bits) are used together to represent integers, then a range of 0 to 65,535 or ($2^{16} - 1$) is possible. For a representation that includes both positive and negative integers, the 2's complement system is most often used. Other representations can be used in which one bit of each number is reserved to represent its sign. Since there are no decimal points in bytes, floating-point representations use an exponential format. A fixed number of bits are used to represent the exponent, and the remaining bits are use for the mantissa. Finally, bytes are used to represent characters, as in word processing. The most commonly used standard code for the English language is the American Standard Code for Information Interchange (ASCII), which defines the bit patterns for 95 printable characters, as shown in Table 1.1. The ASCII standard uses 7 bits for character representations providing 2^7, or 128, unique patterns. There remain 33 bit patterns not used for printable characters; these are used for control codes such as carriage return, backspace, and form-feed. If 8 bits are used, then 256 codes are possible. Many word processing programs use 8-bit code, with special codes for functions such as super- and subscripts.

The second attribute of computer memory is its address, which for each byte is its position in an array of memory bytes. The address uniquely locates each byte in the memory and makes it possible for the CPU to distinguish each byte from the remaining ones. Addresses, like data, are expressed as sequences of binary digits. The length or size of an address varies with different microprocessors. For most of the early microprocessors (e.g., the 6502) addresses were represented by 16 bits, with which 65,536, or 64K, bytes of memory can be addressed. The Intel 8088, which is used in the IBM personal computer (PC), has a 20-bit address bus, allowing for a memory address space of 1000K, or 1 million, bytes.

The addressable space of a microcomputer consists of several kinds of memory cells and often includes input and output devices. Most of the memory of a microcomputer consists of read/write locations often called random access memory (RAM). This type of memory can be both written to and read by the CPU, in contrast to read-only memory (ROM). The CPU has the ability to only read the contents of ROM location. The data stored in ROM memory is considered permanent and is not lost when electric power to the microcomputer is turned off. The contents of the ROM memory are programmed or set by the manufacturer. ROM memory is used to store start-up, or boot, programs in microcomputers and is also used to store commonly used programs, such as subroutines that read and write data to the memory storage system.

In many microcomputers external devices such as printers are connected in such a way that the CPU "sees" them as memory locations. When data is sent to the printer, the CPU writes data to the memory location corresponding to the printer interface. This type of arrangement for interfacing external devices to a microcomputer is called *memory mapping*. In systems that use this technique, sections of memory addresses are set aside for external devices. An example of the arrangement of memory addresses is shown for an IBM PC in figure 1.5.

Disks as Storage Devices

A key component to the growth of microcomputers has been the disk, which serves as a program and data storage unit. The innovation of the floppy disk provided a springboard for the development of microcomputers into powerful multipurpose tools for tasks that varied from word processing to laboratory data acquisition. Later developments such as the compact hard disk, coupled with parallel gains in microprocessor and memory technology advances, have brought about economical desktop computers that rival the computing power of traditional mainframe computing centers.

In the early 1970s IBM introduced the floppy disk or diskette, which consists of a flexible circular piece of thin plastic material coated with an oxide material similar to that used on magnetic tape. When mounted in a drive unit, the disk is rotated at about 360 rpm and data is written and read on the magnetic surface by a head mounted on a movable arm, which moves radially across the rotating disk surface much as does a phonograph arm across the turntable. Data is written on concentric rings called *tracks*, each of which is divided into sections called *sectors*, which store a fixed number of bytes. Typical sector sizes are 128, 256, and 512 bytes. The IBM floppy disk was about 8 in in diameter, and the IBM 3740 single-density format divided this disk into 77 tracks, each in turn divided into 26 sectors of 128 bytes each. This early format allowed for the recording of about 250K bytes of information on a single-sided IBM 3740 formatted disk. An early use of the floppy disk was for the auxiliary storage of data in place of keypunch cards. As the PC developed, the floppy disk was adopted as

TABLE 1.1. The 7-bit American Standard Code for Information Interchange (ASCII) Representation of Alphanumeric Characters, Miscellaneous Symbols, and Terminal Control Codes

ASCII Character	Binary Code	ASCII Character	Binary Code	ASCII Character	Binary Code
NUL	0000000 (0)	+	0101011 (43)	V	1010110 (86)
SOH	0000001 (1)	,	0101100 (44)	W	1010111 (87)
STX	0000010 (2)	-	0101101 (45)	X	1011000 (88)
ETX	0000011 (3)	.	0101110 (46)	Y	1011001 (89)
EOT	0000100 (4)	/	0101111 (47)	Z	1011010 (90)
ENQ	0000101 (5)	0	0110000 (48)	[	1011011 (91)
ACK	0000110 (6)	1	0110001 (49)	\	1011100 (92)
BEL	0000111 (7)	2	0110010 (50)	]	1011101 (93)
BS	0001000 (8)	3	0110011 (51)	^	1011110 (94)
HT	0001001 (9)	4	0110100 (52)	_	1011111 (95)
LF	0001010 (10)	5	0110101 (53)	`	1100000 (96)
VT	0001011 (11)	6	0110110 (54)	a	1100001 (97)
FF	0001100 (12)	7	0110111 (55)	b	1100010 (98)
CR	0001101 (13)	8	0111000 (56)	c	1100011 (99)
SO	0001110 (14)	9	0111001 (57)	d	1100100 (100)
SI	0001111 (15)	:	0111010 (58)	e	1100101 (101)
DLE	0010000 (16)	;	0111011 (59)	f	1100110 (102)
DC1	0010001 (17)	<	0111100 (60)	g	1100111 (103)
DC2	0010010 (18)	=	0111101 (61)	h	1101000 (104)
DC3	0010011 (19)	>	0111110 (62)	i	1101001 (105)

Char	Binary	Dec	Char	Binary	Dec	Char	Binary	Dec
DC4	0010100	(20)	?	0111111	(63)	j	1101010	(106)
NAK	0010101	(21)	@	1000000	(64)	k	1101011	(107)
SYN	0010110	(22)	A	1000001	(65)	l	1101100	(108)
ETB	0010111	(23)	B	1000010	(66)	m	1101101	(109)
CAN	0011000	(24)	C	1000011	(67)	n	1101110	(110)
EM	0011001	(25)	D	1000100	(68)	o	1101111	(111)
SUB	0011010	(26)	E	1000101	(69)	p	1110000	(112)
ESC	0011011	(27)	F	1000110	(70)	q	1110001	(113)
FS	0011100	(28)	G	1000111	(71)	r	1110010	(114)
GS	0011101	(29)	H	1001000	(72)	s	1110011	(115)
RS	0011110	(30)	I	1001001	(73)	t	1110100	(116)
US	0011111	(31)	J	1001010	(74)	u	1110101	(117)
SP	0100000	(32)	K	1001011	(75)	v	1110110	(118)
!	0100001	(33)	L	1001100	(76)	w	1110111	(119)
"	0100010	(34)	M	1001101	(77)	x	1111000	(120)
#	0100011	(35)	N	1001110	(78)	y	1111001	(121)
$	0100100	(36)	O	1001111	(79)	z	1111010	(122)
%	0100101	(37)	P	1010000	(80)	{	1111011	(123)
&	0100110	(38)	Q	1010001	(81)	|	1111100	(124)
'	0100111	(39)	R	1010010	(82)	}	1111101	(125)
(	0101000	(40)	S	1010011	(83)	~	1111110	(126)
)	0101001	(41)	T	1010100	(84)	DEL	1111111	(127)
*	0101010	(42)	U	1010101	(85)			

Characters 0 through 31 and 127 are control characters, such as CR (13), which is a carriage return.

| Address | | Contents | |
Decimal	Hex	Function	Comments
0 16K 32K 48K	00000 04000 08000 0C000	16-64KB RAM installed on System Board	Typically comes with system
64K 624K	10000 9C000	Up to 448KB RAM on option boards installed in I/O expansion slots	Expansion space provides for a maximum of 640K RAM
640K	A0000	Reserved	
656K 688K 704K	A4000 AC000 B0000	Monochrome display buffer	This memory space is used for the terminal display
720K	B4000		
736K	B8000	Color/Graphics display buffer	
752K 768K	BC000 C0000		
 944K	 EC000	192KB memory Expansion area	This space for special software
960K	F0000	Reserved	
976K 1008K	F4000 FC000	48KB system ROM	

FIG. 1.5. A memory map of an IBM PC.

its principal storage device. As disk technology has developed, recording densities have increased and physical size has decreased. The 5¼-in diskette became the standard for the IBM PC generation, and the 3½-in microdisk was first made popular by the Apple MacIntosh and Hewlett-Packard microcomputers. With increased recording densities, 800K bytes can be stored on a double-sided 3½-in microdisk.

Even with increased recording densities of diskettes, rapid changes in the development of 16- and 32-bit microcomputers required larger and faster auxiliary storage devices, which were still relatively inexpensive. One type of hard disk, which has metal disks coated with metal oxide, has met the higher storage and performance demands of the 16- and 32-bit generation of microcomputers. These disk drives typically have several disk surfaces per unit and high rotational speeds (e.g., 3600 rps). Storage capacities also vary from about 5 to 500 megabytes for 5¼-in drives.

The combination of 16- and 32-bit microcomputers with high-performance hard disk drives allow the use of more sophisticated operating systems, on-line storage of large databases, and extraction of information from organized data by powerful statistical and modeling simulation packages normally found only on large computer systems.

INTERFACING WITH THE OUTSIDE WORLD

Within a microcomputer the microprocessor, memory, and auxiliary storage devices are its vital organs. However, without interfaces with the outside world, the microcomputer is of little practical use to its user. Microcomputer interfaces serve as windows or ports for interchanging information with the outside world, which may be a user or laboratory and process control instrumentation. In addition to providing a pathway for data transfer, an interface provides additional functionality for the orderly transfer of data between the world of the microcomputer and a more slowly moving laboratory or industrial environment. From the CPU's viewpoint, interfaces usually are designed to appear as simple memory locations, memory mapped. Each interface has memory locations that serve two functions: as data buffers, through which data are transferred, and as control status registers, which are used to provide information to either the CPU or the external device concerning the status of the data transfer process.

Although many types of external devices may be connected at the non-computer end of an interface, only three types are commonly used in laboratory and industrial environments, namely, parallel, serial, and analog. The parallel and serial interfaces transfer data in digital forms between a microcomputer and an external device (Dessy, 1986c). Analog interfaces, in addition to transferring data, interconvert data between digital and analog forms.

Parallel Interface

The parallel interface resembles the internal bus structure of a microcomputer. Externally, it is a set of parallel wires whose voltages represent the logical states of each wire; internally, it is simply seen as other memory locations by the CPU. One set of wires, usually equal in number to the microcomputer's word size (number of bits), is used to transfer data. The other wires are used for communication of the status of the data transfer process between the CPU and the external device.

Many of the first instruments that used digital electronics provided some type of parallel interface for computer connections (see fig. 1.6). This was often in the form of a connector in the rear of the instrument for a ribbon cable of parallel wires. The electrical states of the data wires or lines represented the output numbers as shown on a digital panel meter. A commonly used meter was a 3½ digit (– 1999 to + 1999), with status lines to indicate overrange and updating or busy state. An updating state is an indication that the instrument is in a measurement state and that the logical states of the data lines during this time do not represent valid data. Likewise, an overrange state indicates the presence of invalid data on the data lines. An often used binary representation of the decimal number on the panel meter is the binary-coded decimal (BCD), which uses 4 bits to represent each number (Table 1.2). For numbers between – 1999 and

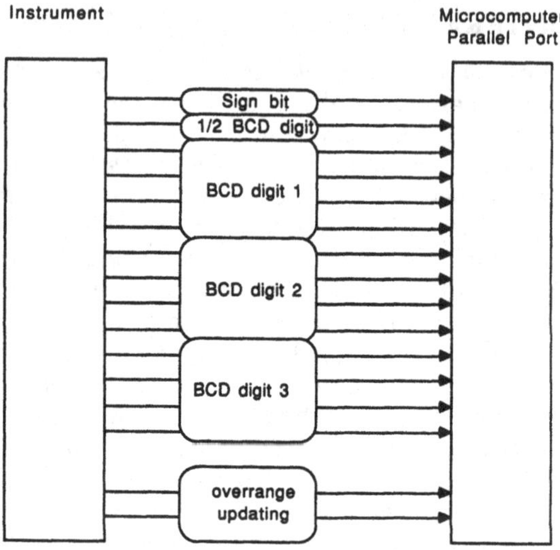

Instrument

Microcomputer
Parallel Port

FIG. 1.6. A parallel interface between a digital instrument and a microcomputer. The instrument is representative of many early digital models, which had a digital display instead of an analog meter and provided a connector through which a BCD representation (Table 1.2) of the digital display could be sent to a computer over a parallel set of conductors.

+1999, 14 bits are necessary, since the first digit ranges from 0 to 1 and 1 bit is used to represent the sign. If a parallel set of wires is to be interfaced with a microcomputer with an 8-bit word-size, then three words would be required for the parallel interface, as shown in figure 1.6; one word would be used for the second and third BCD digits, the next for the sign, ½, and first full BCD digits, and the third for the status lines or bits of overrange and updating state.

Early types of parallel interfaces were subject to very little standardization; the physical type of connector and the assignment and ordering

TABLE 1.2. The Binary-Coded Decimal (BCD) Representation

Decimal	BCD	
0		0000
1		0001
2		0010
3		0011
4		0100
5		0101
6		0110
7		0111
8		1000
9		1001
10	0001	0000
11	0001	0001
12	0001	0010
..	....	....
..	....	....

Many digital panel meters use this binary representation of the decimal digits, which are displayed. With this system, each decimal digit displayed on the meter is represented by 4 bits.

of conductors varied among instrument manufacturers. Likewise, there was little standardization among early parallel interfaces for different microcomputers. One of the first standards to appear on many instruments is known by several names: the IEEE-488 (Institute of Electrical and Electronics Engineers), GPIB (general-purpose interface bus), HPIB (Hewlett-Packard interface bus), and ANSI MC1.1 protocol (American National Standards Institute). This standard specifies eight lines for data, eight for status control and handshaking, and eight ground lines (Dessy, 1986c) and also specifies that a 24-conductor cable be used with a special type of connector. In addition to standardization the IEEE-488 interface has a higher level of intelligence than the generic parallel interface discussed above. The latter type of interface is for the connection of one external device to a microcomputer, whereas the IEEE-488 standard allows for connection of up to 15 instruments to the same interface on one microcomputer. Each instrument has a different address on the interface, much like the address structure inside a microcomputer that is used by the CPU to address different memory locations. Thus the IEEE-488 parallel interface is often called an *instrument bus*.

Another type of intelligent parallel interface, which appeared on the Apple MacIntosh Plus microcomputer, is the small computer system interface (SCSI, ANSI document X3T9.2/82-2). This interface was first designed for the connection of different types of auxiliary storage devices to the same microcomputer through a single interface. Being highly intelligent, the SCSI interface relieves the CPU from dealing with the complexity of data transfers among several different types of devices. The SCSI host adapter is an interface between the internal bus of the microcomputer and the SCSI bus, to which the different types of storage devices may be connected with an appropriate controller or interface, as shown in figure 1.7. As a general-purpose bus, the SCSI can support more than mass storage devices; laboratory instruments with an SCSI controller could also be connected to this bus. With its ability to support multiple hosts, an SCSI bus could interconnect multiple CPUs into a local area network; for example, in a local laboratory environment it could interconnect instruments, microcomputers, and mass storage devices.

Serial Interface

A second interface, which also transports digital information between a microcomputer and an external device, is called a *serial interface*. In parallel data transmission each bit of a word has its own conductive path; in serial transmission all data bits share one conductor and are sent one at a time on this conductor. A separate conductor is used for data transmission between the two devices in the reverse direction. The obvious advantage to this is the reduction in cabling needed to complete an interface, which would suggest that serial interfacing would be the practical choice when an external device is remotely separated from a microcomputer. At long

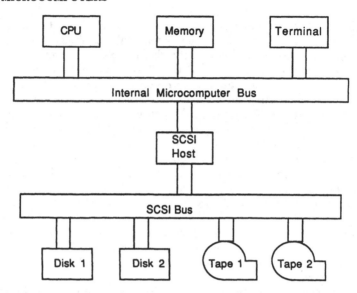

FIG. 1.7. A small computer systems interface (SCSI) to a microcomputer simplifies the addition of different types of mass storage devices and still provides high-performance data transfers.

distances and high transmission rates, copper conductors begin to distort signals because of capacitive effects. In order to ensure accurate transmissions over long distances, the rate of transmission is often slowed to prevent electrical distortions of digital bits. *Baud rate* is a term often used to specify the transmission rate in bits per second (bauds); commonly used baud rates are 300, 1200, 2400, 4800, 9600, and 19,200 bauds. As new conductive paths, such as optical fibers or microwave links, or other innovations in electronics are developed, higher baud rates are becoming available for high-volume data transmissions.

For serial transmissions in laboratory environments, most devices such as terminals and instruments follow a standard known as the RS-232C (Electronics Industry Association) or the V.24 (Consultative Committee on International Telegraphy and Telephony). Although not part of the standard, the data code most commonly used in serial transmission is the ASCII code (Table 1.1). Each transmitted character is prefixed with a start bit (logic 0) and appended with one or two stop bits (logic 1) (fig. 1.8). The function of the start bit is to signal the receiving end that the beginning of a character is following with the next bit. In the dormant state the line is held in a logic 1 state so that the change to logic 0 represents this signal. The stop bit gives the receiving end an opportunity for a pause before another character is received. An optional bit may be added before the stop bit(s) for parity or error checking. Odd or even parity is invoked by writing a logic 1 or 0 in this bit so as to force the number of 1 bits to be odd or

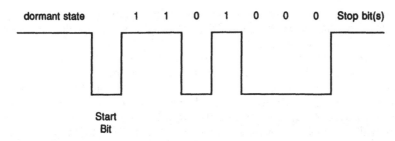

FIG. 1.8.   A serial transmission of the 7-bit string 1101000, which is
the ASCII representation of the character "h" (Table 1.1).

even. The receiving end of the character can check the parity bit and number of 1s as a limited check for transmission errors.

The RS-232C standard also provides for the optional addition of other lines that may be used to coordinate the serial transfer of data. Although many instruments, printers, terminals, and other devices use the RS-232C interface, their implementation of the various control lines may vary. Some devices use none of the control lines but only the data transmit and receive lines with a ground line. Even though many microcomputers and devices have RS-232C serial-type interfaces, there is still room for incompatibility. Each device must recognize and respond to the contol lines used by the other device. Also, the form of the data sent by the transmitter must be the same as that which the receiver end expects. The baud rates must be set the same, and if parity is used, then it must be enabled in the same manner (even or odd) at both ends.

Analog Interfaces

Analog interfaces are usually required in computer-based data acquisition systems. Many basic phenomena are studied by the measurement of related physical or chemical parameters, such as temperatures, light absorbance, and pH. The measurement process often involves conversion of a physical parameter into an electrical signal, which represents the magnitude of the parameter is represented by the magnitude of the electrical signal. Analog electrical signals include all electrical quantities (such as voltage, current, resistance, and capacitance) whose magnitude is related by a continuous function to a physical parameter (Malstadt and Enke, 1969).

When a microcomputer is used for data acquisition, several data conversions are often necessary. For example, measuring instruments may contain transducers, which convert physical data into analog signals. A thermocouple converts temperature into an analog voltage. A further conversion from analog to digital is necessary before microcomputer data acquisition is possible. For microcomputer control of laboratory instruments, digital-to-analog signal conversion is often required. The intensity of a heater of a gas chromatograph oven could be controlled by the magnitude of an analog voltage or current. If temperature control by a microcomputer is desired, then a conversion of digital numbers into analog voltages for the heater is necessary.

Microcomputers can send and receive only digital signals. They communicate with the external world through interfaces. Parallel and serial interfaces are used for connection to devices that are built with some digital electronic circuitry. Today most laboratory and process monitoring and control instrumentation is built with digital circuitry with a standard serial or parallel interface. However, a good many older equipment systems, which have no digital interfaces but were built solely with analog electrical systems, are still in use. Also, many scientists and engineers design and build their own measurement devices with analog transducer systems. In either case analog-to-digital conversion is necessary for interfacing to a microcomputer.

Analog voltages are converted to digital signals with devices called analog-to-digital converters (ADCs), and digital signals are converted to analog voltages with units called digital-to-analog converters (DACs). Analog-to-digital conversion in its basic conceptual form involves two steps, quantizing and coding (Zuch, 1979)—*quantizing* is the process of transforming an analog signal into an ordered set of discrete binary digits, and *coding* is the assigning of a digital value to each of the possible states. A detailed example of these different aspects of analog conversion is presented in Chapter 2.

SOFTWARE

A microcomputer has two distinct functional systems components, hardware and software. The hardware components are the physical machine parts of the system, including the microprocessor, memory, bus, I/O and peripherals. The second component of a functional system is software, that is, computer programs. Modern microcomputer hardware is very affordable and provides desktop computing power comparable with that of the corporate and academic mainframe computer centers of the 1960s. Advances in solid-state technologies and auxiliary storage devices have brought about these rapid changes. However, without the software this equipment is little more than an inanimate collection of advanced solid-state circuits.

The driving element that harnesses the power of the hardware is the software. A computer program is a series of computer instructions organized to produce some meaningful manipulation or transformation of input data into output data. Computer programs can be classified in a number of different ways. For microcomputers, there are three ways to distinguish levels of software with regard to functionality. *Application software* consists of programs that perform particular functions, such as word processing. *Programming languages* are software "rules of discipline" with which programmers develop new application programs. *Operating systems* are collections of programs that allow users to use application programs or to create their own programs with a given programming language.

Application Software

The rapid integration of microcomputers into various aspects of our existence has happened because of the low cost of hardware and application software. This has been a reflection of the fact that production of good software is a very difficult task, even for experienced programmers. Thus, before the computer could become a useful tool to many, the existence of prewritten programs that performed universally useful tasks was necessary. The first and still most widely encountered task for which application software was successfully written has been word processing. Within the short span of several years, the typewriter has been largely displaced by the development of word processing software for the microcomputer. Word processing software is a generalized software package, as it is used by secretaries for document preparation and by computer programmers to prepare programs in a programming language.

Two other application programs that followed the success of word processing programs were the electronic spreadsheet and the database manager, both of which allow the user to develop a varied number of procedures for specialized analysis of data. For example, a data base management program allows a user to define the setup of the data records, including field name, length, and kind of data to be stored in each. After a data base is created, the user can use the program to retrieve records based on programmable criteria and select a format for display of the data. Such application programs take on aspects of a programming language but are not classified as such since they limit the resources of the computer to which the user has access. The advantage of these application programs is that their learning curve is much shorter than that for a new programming language. Some call such application programs "programming languages of a higher order."

For the laboratory there are a number of application programs, many of which combine a number of specialized functions into a scientific format. One such program is RS/1 (Bolt Beranek and Newman Inc., Cambridge, MA). This program is recognized as an electronic notebook for the laboratory as it provides data storage, statistical analysis, and graphics capabilities in one program. Many types of electronic notebook programs have been developed; increases in their power and ease of use have followed closely on the development of more powerful microcomputers. For most users programs of this type are the most important, since most scientists and engineers do not have the time to become expert programmers.

New concepts in application programs include expert systems and artificial intelligence. *Expert systems*, by their nature, are very specialized but powerful programs or systems. Some general aspects of these systems are that they derive their knowledge and reasoning abilities from human experts and possess most of the following attributes:

They cover a specific domain of expertise.
They contain a knowledge base that is organized as a collection of rules rather than hardcoded into the deductive process.

They can be used to reason with uncertain data and reveal the results of that reasoning in an understandable way.

They can grow incrementally.

They deliver advice rather than charts or figures.

The distillation of the experts' knowledge into machine-readable language is a challenge both to experts in the process under consideration and to expert programmers.

Programming Languages

All computer software is written by programmers in some programming language. The development of programming languages, like that of computing hardware, has progressed through several steps, or generations.

A first-generation language is the machine code closest to the CPU architecture. A program statement is a string of binary 0s and 1s, which directly controls the maze of logic gates within a CPU. All operations, including arithmetic ones, are done in binary, and how a program is coded directly reflects how operations are to be performed by the machine.

Second-generation languages, or assembly languages, brought about the substitution of names for binary strings. The names were chosen so that they reflected the action of the machine instruction: for example, ADD for addition and MOV for moving data from one memory location to another. Symbolic names were substituted for actual addresses in memory when possible. The assembly language program did the tedious, error-prone work of converting the symbolic addresses and instruction names into binary machine code. A simple operation still took many discrete steps, however, and decisions were made according to condition tests, that is, according to the state of the machine's registers, instead of the task's inherent logic.

The third-generation languages gave statements that had forms closer to the task to be accomplished than to the architecture of the microcomputer on which it was to run. Assembly languages are dedicated to the specific microprocessor. Since third-level languages were designed to program tasks rather than microprocessors, their form could be standardized without regard to which microprocessor is used. High-level program statements usually corresponded to large numbers of machine instructions, and choices of action based on logical conditions replaced the comparisons of register contents and condition tests, thus moving closer to human perception of the task to be done and away from the machine's demands for a binary representation.

In technical areas FORTRAN is the oldest high-level language that is still widely used. It is over 30 years old (having been developed in the 1950s) and has undergone several major revisions. It has been reputed to be a dying language more than once, but in many technical fields it is still the language most scientists and engineers use for routine programming for numerical analysis. There is considerable debate as to why FORTRAN is still so popular, but whatever the answer, it appears that FORTRAN still has a bright future for numerical calculations in technical fields.

With the introduction of first-generation microcomputers, which had limited memory and auxiliary storage devices, the BASIC language became a standard for these models. Because of the simplicity of this language, it was much easier to implement on the limited resources of these early models. As microcomputer hardware became more advanced, the popularity of FORTRAN returned, along with other newer languages such as PASCAL and the C language.

The fourth generation arrived in the form of several new languages. Historically, languages have moved farther away from the machine level, becoming more abstract with each generation. The fourth generation can be thought of as crossing the threshold into a world where the programmer specifies the task to be done and the knowledge of how to do the task is contained in the language itself. LISP and PROLOG are two new languages which are most often mentioned as fourth-generation. Two earlier languages that are sometimes placed in the fourth generation are APL and FORTH.

APL was an early attempt to program according to the logic of the problem rather than to the architecture of the machine. Originally a notation for applied mathematical algorithms, APL has been adopted by IBM and Digital Equipment Corp. and is available on many 32-bit microcomputers. FORTH is a compact language whose original popularity was due to its use with small microcomputers. It is also very interchangeable among computers and is very popular as an advanced language for developing interfacing applications between microcomputers and instruments. Advanced application programs are sometimes included in the fourth generation, for example, with some data base management systems.

The often stated goal in developing new generations of languages is to have them accept English-like sentences as input and perform the tasks that these sentences describe. Fourth-generation languages are not as much like written English as many users expect, but they are powerful and easy to use. The trend toward natural-language software is continuing.

Operating Systems

In order to use application software or programming languages on a microcomputer, an operating system is necessary. An operating system is a collection of programs that provides for the start-up of the microcomputer on power-up; the loading, executing, and storage of programs; storage and retrieval of files; and the execution of utility programs. The operating system also provides for communication between the user and the system and between the various hardware devices and the system.

Each device interfaced to the microcomputer normally has a program called a *driver*, or *handler*, as part of the operating system. Data transfers to and from the device are controlled by this driver program. When application or user programs developed with a programming language require external data transfers, they pass the request to the operating system, which calls the appropriate driver program. The use of these driver programs greatly simplifies the development of new programs since requests for input or output of data are passed to the operating system. When a new device is added to a microcomputer, a new driver program must be written and added to the operating system.

The first operating system developed for personal computers was CP/M. AppleDOS (Disk Operating System) and Apple ProDOS are two operating systems that are used for the earlier versions of Apple II computers. With the advent of the IBM PC, MS-DOS or PC-DOS became standard for the Intel 8088 microprocessor class of microcomputers. Earlier operating systems, first developed for Digital Equipment's PDP-11 minicomputer series and later adapted to its LSI 11 series of 16-bit microprocessors, were RT-11 and later RSX-11M. These operating systems were designed to provide real-time response of the microcomputer to external events and to enable the microprocessor to execute multiple tasks concurrently. Although the CPU can only execute one program at a time, it can rapidly alternate between tasks with such speed that it appears that the different tasks are being executed simultaneously. In a real-time system priorities are assigned to each task and associated device. When a task or its device requests action from the CPU, the CPU execution of the current program is interrupted. If the interrupting task has a higher priority than the program being executed, the CPU sets aside the current program and executes a program to service the interrupting task or device. If the interrupt is of a lower priority, it is set aside until the higher-priority task is completed.

Operating systems, like programming languages, are always being upgraded. Also as with programming languages, there is a drive to make operating systems respond at higher levels to make their use more "user friendly." One true innovation is the Apple MacIntosh operating system, which has jumped over the English-like interface and gone to a graphic interface with icons or symbols used to represent programs and files and a "mouse" used to point to symbols instead of typing names on a keyboard. Many of the 32-bit work stations are being developed with this type of user interface within their operating systems.

REFERENCES

DESSY, R. E.   1986a.   Choosing a PC. Part I. Anal. Chem. 58:78A–91A.
DESSY, R. E.   1986b.   Choosing a PC. Part II. Anal. Chem. 58:313A–333A.
DESSY, R. E.   1986c.   The PC connection. Part I. Anal. Chem. 58:678A–689A.
MALSTADT, H. V., and C. G. ENKE.   1968.   Digital Electronics for Scientists. New York: W. A. Benjamin, pp. 280–288.
ZUCH, E. L.   1979.   Principles of data acquisition and conversion. In E. L. Zuch (Ed.) Data Acquisition and Conversion Handbook. Mansfield MA: Datel Intersil, pp. 1–26.

2

Data Acquisition in the Laboratory

Analytical chemistry may be defined as the science and art of determining the composition of materials in terms of the elements or compounds that they contain (Ewing, 1985). The role of the computer in aiding this endeavor has evolved over three stages, namely, data processing, data acquisition, and data communications. The first use of the computer in analytical laboratories was for data processing, principally for statistical analysis of data on remote mainframe computers. During this period computer technology had little immediate impact within the laboratory. The initial analysis of raw data, which included measurement of distances or lengths associated with experimental parameters, was still a manual operation in the laboratory. The form of raw data varied from the displacement of a pointer on an analytical balance to a pen that traces by its vertical movement on horizontally moving chart paper. With the advent of the minicomputer in the 1960s, the vigorous development of computerized data acquisition in the laboratory began. The electronic capture of data made possible elimination of many manual manipulations, and consequently a larger number of analyses were possible. These developments brought about the need for electronic pathways to transfer data from dedicated microcomputers to larger computer systems for more extensive data analysis, report generation, and data storage.

Although a chronological order exists with regard to the three stages, all have undergone further development. For example, developments in artificial intelligence and expert systems indeed represent higher orders of data processing. Data acquisition has grown from data capture through instrument control into laboratory robotic systems. Current levels of automation have resulted in electronic laboratories, where a collection of intelligent instruments coexist with a number of microcomputers and supermicrocomputers. The effective management of such laboratories has brought about electronic networks, or local area networks (LAN), over which information can be shared. The control of such networks is often maintained by laboratory information management systems (LIMS), and the sophistication of these communications systems is continually growing. The cumulative potential of these developments in food analysis laboratories is bringing about tremendous advances in the size of workloads and the management and quality of data produced.

THE MEASUREMENT PROCESS

In the analytical laboratory, the digital computer is a ubiquitous machine, the application of which originated in the measurement process. The transformation of data from a physically measured quantity into an electronic digital signal serves as the foundation for the electronic laboratory. Because of the importance of this process its fundamental principles should be understood by analytical scientists.

It is the function of an instrument to translate physical or chemical characteristics of a material into a form directly observable by the operator. Most measurements made in the laboratory have consisted of observations of linear or angular displacements evaluated by comparison with some kind of scale. Even when electronics first became important to chemists with the introduction of the glass electrode for pH measurement in the 1930s, distance measurement was still the final form in which information was transferred to the operator by the displacement of a pointer on an electronic pH meter.

However, the advent of electronics did bring about the development of a new generation of instruments. Advances were primarily centered around electronic transducers, which were devices that translated or converted physical or chemical information into an electronic form. For example, the photoelectric cell converted the magnitude of the intensity of incident radiation to an electric current of proportional magnitude. A major advantage of electronic transducers was that circuitry could be added to amplify the electronic signal. In this way analytical instruments were designed to be as sensitive as practicable, so that they were able to measure precisely the smallest signal that could be produced by the transducer. Instruments of this generation included a second transducer to convert the electronic signal to an operator-observable form, which varied from a meter displacement to a pen movement on a strip chart recorder. Electronic transducer systems in analytical instruments have brought about improvements in detection limits, the ability to directly measure new properties, and the ability to continuously record the magnitude of parameters during an experiment. Analytical instrumentation continued to improve as new advances in electronics were made.

One significant point of development was the advent of the transistor and other semiconductor devices. Solid-state electronics served to increase the sensitivity of measurements and the development of new transducers. However, solid-state electronics will be primarily noted as the foundation for the development of the digital computer. Likewise, the further development of integrated solid-state circuitry brought about the introduction of the microprocessor, which has in turn brought about a new generation of analytical instruments.

The innovation represented by the small inexpensive microprocessor has resulted in its inclusion in almost every modern electronic instrument. The microprocessor is used for two functions in analytical instruments, namely as a control element and as a data recorder. Microprocessor control and optimization of instrument operation is a complex task. Its implementation is usually carried out by electrical engineers, and the resulting control system is usually transparent to the scientist. Although the data acquisition part of an instrument may also be transparent, its effective use often requires a basic understanding of digital data acquisition. Many instrument manufacturers provide the user with the ability to program the data acquisition component to be used by the instrument's microprocessor during an analysis. This flexibility greatly expands the instrument's utility but also requires a more knowledgeable user. In order to identify the role of a number of data acquisition parameters in a digital system, a general case will be considered.

GAS CHROMATOGRAPHY: A CASE STUDY

An instrument universally used in analytical laboratories is the gas chromatograph (GC); gas chromatography is one of the most extensively used analytical separation techniques. A schematic diagram of a typical GC with data system is shown in figure 2.1. A mixture of compounds is introduced through the sample injector into a steady stream of carrier gas. As the mixture is transported through the column, which contains a stationary phase, a separation process occurs. The separation of the mixture's

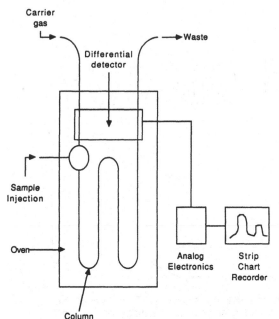

FIG. 2.1. Schematic diagram of a typical gas chromatograph. (From Ewing, 1985.)

components occurs by continuous partitioning of each component between the carrier gas, the mobile phase, and the stationary phase within the column. The partitioning ratio is normally different for each component, which leads to a separation of the components of a mixture as they travel through a column. The GC is equipped with a detector, whose function is to quantitatively detect the presence of compounds as they leave the column. The heart of the detector is a transducer, which produces an electrical signal of a magnitude proportional to the concentration of an eluted component. Production of an observable record, that is, a *chromatogram*, of this separation process requires another transducer, which must convert the electrical signal produced within the detector to a physical form that is readable by the operator and convenient for numerical analysis. Before the advent of the computer, the universal converter system was the strip chart recorder. A strip chart recorder (fig. 2.2) is built with a servomechanism, which produces a mechanical or physical effect proportional to a varying signal. In this case the physical effect is the movement of a pen with a displacement proportional to a voltage produced by the transducer within the detector. The strip chart recorder produces a visual or graphic record of the GC analysis. In this form the recording is suitable for further analysis by the operator. A gas chromatogram appears as a series of peaks (fig. 2.3), each of which represents various kinds of information. The position of a peak relative to the start of an analysis is related to the identity of a compound, and the area under the peak is related to its concentration. The shape of the peak indicates other information concerning the analysis, including column type, carrier gas flow rate, and column temperature. The mea-

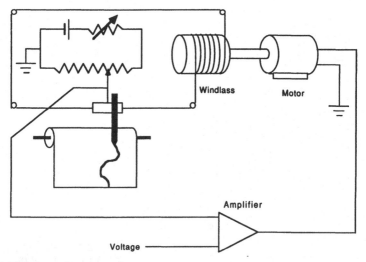

FIG. 2.2.   Servo system for a strip chart recorder. The motor-driven carriage moves both the sliding contact on a voltage divider and the recording pen. (From Arnold, 1979.)

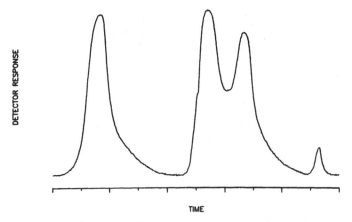

FIG. 2.3. A recording of an analog signal from a gas chromatograph.

surements and calculations necessary to convert the graphic record into a numerical record are straightforward but tedious, particularly when the number of components in a mixture or the number of analyses conducted is large, both of which situations are common in food analysis laboratories.

The ability to generate large masses of graphic data in relatively short periods of time has been a problem inherent in gas chromatography as well as with other modern instrumentation. The advent of the microprocessor has provided a cost-effective building block for automated data collection and analysis. A computer can easily perform the calculations, but translation of the data from a strip chart recorder and its manual entry into a computer often constituted the more time-limiting task. In order to remove this bottleneck the computer must electronically capture the data directly from the instrument. Since the signal is usually in analog form and computer signals are digital, a conversion process called analog-to-digital (A/D) conversion is necessary.

Devices that perform A/D conversions are called *analog-to-digital converters* (ADCs). The choice of an ADC subsystem as an interface between an instrument and computer demands careful analysis; otherwise, the interface may create data rather than simply changing its form from analog to digital. To avoid the corruption of data during its conversion, it is necessary to appreciate the relationship between data in analog and digital forms and the adjustable parameters of the process.

The effect of digitizing an analog signal is illustrated by comparing figure 2.3 with figure 2.4. Whereas an analog signal is a set of continuous points, infinite in number, the digitized signal is a set of a discrete number of points. The number of points in the digital signal should be adequate to maintain a sufficient level of precision in the signal. However, there may be experimental conditions that prevent the acquisition of an adequate number of points. These conditions are formed by the interactions present at the interfaces between the analog signal and the digital computer.

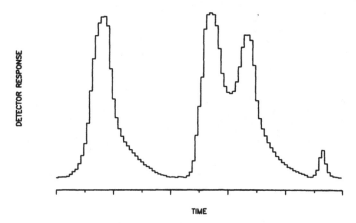

FIG. 2.4.   Digital recording of the analog signal shown in figure 2.3.

Interface Hardware

The first hardware parameters to be considered are those of the ADC subsystem itself; these include A/D conversion time, voltage resolution, and the components other than the ADC unit in the subsystem. A proper match between these parameters and the event to be measured is necessary for good measurements. For experiments of short duration, the time required for each A/D conversion may be a limiting factor. The effect of a relatively slow sampling rate is illustrated in figure 2.5. The loss of information can be appreciated by comparing this figure with the original analog signal in figure 2.3. Experimental data such as peak centers, shapes, and areas can be distorted by the loss of resolution of time. Also, identification and separation of overlapping peaks becomes more difficult.

The conversion time of an ADC depends strongly on its type. Two major types are used with data acquisition computers in analytical laboratories; namely, the integrating and the successive approximation types (Dessy, 1986). The integrating types, which are normally dual-slope integrating ADCs or voltage-to-frequency ADCs, are noted for their ability to filter out high-frequency noise from the data signal by electronic integration techniques. Their limitation is that they have relatively slow conversion rates, varying from 60 to 0.1 samples per second (hertz). The limitation of 60 Hz comes from the frequency of the most common electronic noise contaminants in laboratory environments. When faster conversion times are needed, successive approximation ADCs are used; these have typical conversion rates of 30,000 samples per second, or 30 kHz, but they have the disadvantage that high-frequency noise is carried through the conversion process and into the digital data set. The effects of noise on data signals will be discussed later.

The other important parameter of an ADC is its steep *height*, determined by dividing its analog input range by the total number of digital resolution

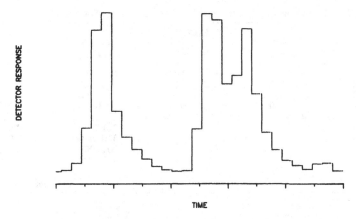

FIG. 2.5.   A low-resolution digital recording of the analog signal in figure 2.3, obtained with a slow sampling rate.

steps that it can provide. Typical analog input ranges, or full-scale (FS) voltages, are −10 to +10 V, −5 to +5 V, 0 to +10 V, and 0 to +5 V, and for most ADC systems the range can be selected from among these. The digital resolution is determined by the number of output bits. Common values used in analytical instrumentation are 10 to 16 bits; these provide for 1024 (1K) to 64K possible output states, respectively. For an ADC with an FS voltage of 0 to 10 V and 10 bits of resolution, the step size would be 10 V per 1024 bits, or 9.77 mV per bit. The assignment of the relationship between the analog voltage input and the binary-coded output may vary between ADCs and is set by the manufacturer. Some commonly used codes are illustrated in Table 2.1. The step size of an ADC may be thought of as the vertical analog of the sampling time, which may be considered the horizontal step size.

The ADC is the heart of the A/D interface; however, it is seldom the only unit in such interfaces. In computer data acquisition systems, functional ADC subsystems usually include several devices that are placed between the analog source and the ADC, including amplifiers, filters, and multiplexers (fig. 2.6).

Although most ADCs provide some user selection on the FS voltage range, many transducer systems produce FS readings far below the most sensitive range of common ADCs. For example, to measure temperature variations between 0 and 100°C, the FS voltage range of a common thermocouple is about 4 mV. The 10-bit ADC discussed above, with an FS voltage of 10 V, would provide no resolution, since its step size is 9.77 mV. Instrument amplifiers are commonly used to provide a better match between the analog transducer and the ADC. Some ADC units are provided with instrument amplifiers with which the gain can be changed by selection of an appropriate resistor, whereas in some units software-programmable or autoranging gains provide the capability for the computer to select the gain.

TABLE 2.1. Typical Binary Codes Used in Commercial ADC Subsystems

Scale	Straight Binary	Binary-Coded Decimal		Offset Binary	2's Complement
+FS	11111111	1001	1001	11111111	01111111
+¾ FS	11000000	0111	0101	11100000	01100000
+½ FS	10000000	0101	0000	11000000	01000000
+¼ FS	01000000	0010	0101	10100000	00100000
0	00000000	0000	0000	10000000	00000000
-¼ FS				01100000	11100000
-½ FS				01000000	11000000
-¾ FS				00100000	10100000
-FS				00000000	10000000

|_____Unipolar_____| |_____Bipolar_____|

2's Complement	Sign Extended
01100000	0000000001100000
10100000	1111111110100000

Offset binary is a shifted straight binary code where one-half the full-scale binary number corresponds to analog 0. 2's complement coding is related to offset binary by a complementing of the most significant bit. BCD can be converted to straight binary by multiplying each base-10 digit by its corresponding binary equivalent and summing the results. Sign extension involves taking the most significant bit of the ADC output, which represents the sign (0 = +, 1 = -), and duplicating it to the left to the limit of the computer's word length. Extension is necessary for subsequent data manipulations. FS = full scale voltage.

Another device commonly included on an ADC subsystem is an analog multiplexer. Often there is more than one instrument in the vicinity of a laboratory computer, but only one ADC is available. This requires some type of switch by which the computer can acquire data from the different sources by switching the appropriate analog signal to the ADC. Rather than using a mechanical switch, ADC subsystems have built-in multiplexers, which are solid-state switches and are computer-controlled. A multiplexer on the front end of an ADC can be connected to a number of analog signals from various instruments, and selection of the analog source will be made by the program before it begins the acquisition of data. If a programmable gain amplifier is also present in the subsystem, the program can also select the gain factor.

A second set of important parameters that affect the A/D conversion of an analog signal is related to the interface between the ADC and computer and to the software that controls data transfer across this interface. This set includes the method of starting (triggering) each data conversion by the ADC and the rate at which the computer can retrieve and store data from the ADC subsystem.

An ADC has two principal control lines, the start, or trigger (TRIG), line and the end-of-conversion (EOC) line. The start line is a digital input line to the ADC. A change of logic states on this line will initiate an A/D conversion. The specification of the protocol of the necessary logic change (0 to 1 or 1 to 0) varies among different makes of ADC units. Since the

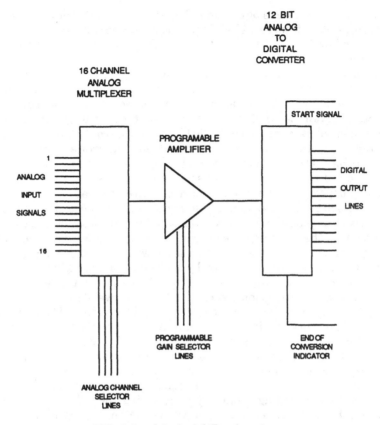

FIG. 2.6.   A typical A/D subsystem.

duration of the conversion process also varies among types of ADC, the EOC control line is used by the ADC to signal the completion of a conversion by a change of logic on this line.

The digitization of an analog signal produces a set of discrete numbers, each of which represents the magnitude of the signal at some point in time (see figs. 2.3 and 2.4). The value of this set of numbers is considerably lessened without the knowledge of the corresponding time coordinates. Normally, the time interval between conversions is constant during a data capture process; thus, knowledge of the interval and the position of a point in the captured array specifies the time coordinate of each point. There are usually three ways by which the rate of A/D conversions is controlled during an experiment—programmed control, external triggering, or an internal clock in the ADC subsystem. Under programmed control the computer starts each conversion after the elapse of each sampling interval created by software. In addition to acquiring and storing data, this method requires the computer to keep track of the time. With fast ADCs, the effective sampling rate is often limited by the computer's ability to retrieve and process data from the ADC. Thus, the sampling rate is often controlled by an electronic clock, either in the ADC subsystem or external to the computer system.

An electronic clock is a circuit that produces digital pulses at a fixed rate. When connected to the TRIG input, this series of pulses would initiate A/D conversions at the same rate as the clock. Many ADC subsystems provide programmable electronic clocks. With this configuration the rate of A/D conversions can be controlled by the same computer program that controls acquisition of data from the subsystem. The choice of the conversion rate during A/D conversion of an analog signal is a critical parameter. The resolution of the signal and the separation of data from noise are both strongly dependent on a proper choice of conversion rate.

A common problem to most instrument signals is noise. This noise may be specified as peak-to-peak noise, which is measured from the most positive excursion to the most negative excursion on the signal. The effect of the addition of noise to an analog signal is shown in figure 2.7. The degree of noise in a data signal is specified by the signal-to-noise (S/N) ratio.

The removal of extraneous noise from the data signal is an important part of any data acquisition process. Fast ADCs have little noise immunity and often require filters to remove noise from the signal. The band width of any signal must be limited, or high-frequency noise may be aliased to lower frequencies in the digitization process, which can cause further corruption of the data. In order to select proper filter units, knowledge of the frequencies of the noise and data components of the signal is necessary; this may require observation of the signal with a device such as an oscilloscope. If the sampling frequency used during the A/D process is lower than the highest-frequency components of the analog signal, an *aliasing*

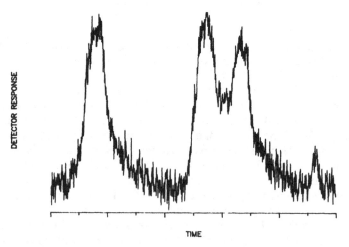

FIG. 2.7.   A noisy version of the analog gas chromatograph signal of figure 2.3.

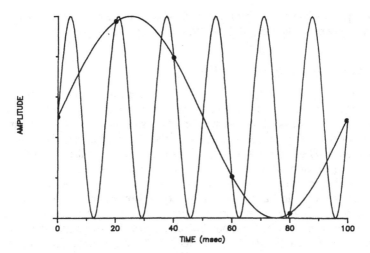

FIG. 2.8.   An analog signal of a 60-Hz sine wave sampled at a rate of 50 Hz. The result is the aliasing of the 60-Hz signal to a signal with a lower frequency.

effect can occur (fig. 2.8). The aliasing of high-frequency signals to lower frequency applies to both noise and data components of the composite signal. The aliasing of the noise in the chromatogram of figure 2.7 to lower frequencies by the use of a slow A/D conversion rate is illustrated in figure 2.9. Filtering of the signal may be done with the digitized data by various mathematical algorithms, but if the aliasing of high-frequency noise into

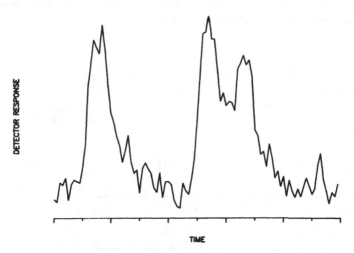

FIG. 2.9.   The aliasing effect of sampling the noisy analog gas chromatograph signal shown in figure 2.7 with a sampling rate of lower frequency than the noise.

the frequency band width of the data signal occurred during acquisition, the noise cannot be removed by programmed techniques.

Software for Data Acquisition

The other essential component for digital data acquisition, in addition to interface hardware, is software, or a computer program to control the data capture process. This software could also provide additional functions, including filtering of noise, data analysis, and identification of sample components. The program code for data acquisition is often straightforward for the acquisition of an analog signal. For example, a BASIC program for a personal computer (PC) with an ADC subsystem is listed in figure 2.10. This program could be used for the acquisition of the gas chromatogram in figure 2.3. However, one experimental concept that is not built into the program is synchronization, so that the start of data acquisition must be synchronized with the start of the GC run. This synchronization can range from the typing of the command RUN on the computer keyboard at the same instant as the sample is injected into the GC column to more elaborate electronic synchronization between instrument and computer.

When the analog signal looks more like figure 2.7 than figure 2.3, noise can be removed from the data signal by digital programming techniques, provided that the frequency components of data and noise are adequately separated and that aliasing did not occur during the acquisition process (fig. 2.9). The separation of noise from data is called *data filtering* or *smoothing*. Many mathematical theories and techniques are used to achieve noise removal, and their application has provided better detection limits in many types of analytical instruments. Four common techniques used for filtering are boxcar averaging, weighted digital filtering, ensemble averaging, and frequency domain analysis.

```
10 'THIS IS A PARTIAL LISTING OF A BASIC PROGRAM FOR AN IBM PC
20 'WITH AN ADC BOARD (DATA TRANSLATION INC., MARLBOROUGH, MA,
30 'MODEL DT2814). THIS SECTION ACQUIRES 1000 POINTS AND STORES THEM
40 'IN AN ARRAY 'Y'. THE RATE OF A/D CONVERSIONS IS CONTROLLED BY A CLOCK
50 'ON THE ADC BOARD, AND THE ANALOG GC SIGNAL HAS BEEN CONNECTED TO CHANNEL
60 'ZERO OF THE ADC BOARD.

100 DIM Y(1000)                      'Y IS AN ARRAY TO STORE THE POINTS
110 OUTPUT &H220,16                  'INITIALIZES AND STARTS THE ADC
120 FOR I=1,1000                     'SETUP AND START LOOP FOR ACQUISITION
130 A=INPUT(&H220)                   'CHECK ADC TO SEE IF NEXT CONVERSION DONE
140 IF A =128 THEN 150 ELSE 130      'IF DONE READ DATA IF NOT CHECK AGAIN
150 B=INPUT(&H221)                   'GET HIGHER ORDER BYTE (BASE 16)
160 C=INPUT(&H221)                   'GET LOWER ORDER BYTE
180 Y(I)=256*B+C                     'COMBINE INTO ONE NUMBER (DECIMAL BASE)
190 NEXT I                           'CONTINUE UNTIL 1000 POINTS RECORDED
```

FIG. 2.10. A partial listing of a BASIC program for an IBM PC equipped with an ADC subsystem for acquisition of an analog signal.

Boxcar averaging is a simple and widely used averaging technique which averages a sequential set of data points and uses the average to represent the data value for the time interval span over which the points were measured. This is based on two assumptions, namely, that the magnitude of the analog signal varies slowly with respect to the sampling rate of the ADC and that the average of a small number of points will be a better measure of the signal than any one of the points. The technique of boxcar averaging can be illustrated by its application to the noisy analog GC signal of figure 2.7. If this signal is sampled with an ADC at a rate that divides the signal into 1000 intervals, the resulting set of data points could be smoothed by boxcar averaging. For example, the set of 1000 points can be subdivided into 100 subsets of 10 points each, and the value of the analog signal for each subset would then be approximated by the average of the 10 points of the subset. The original set of points is represented by y_i, where $i = 1, 2, \ldots, 1000$, and the smoothed set of 100 points is represented by

$$y'_j = \left(\sum_{k=m}^{n} y_k \right)/10 \tag{2.1}$$

where

$$j = 1, 2, 3, \ldots, 100$$
$$m = (j - 1) \cdot 10 + 1$$
$$n = j \cdot 10$$

The results of boxcar averaging with the above parameters on the analog signal in figure 2.7 is illustrated in figure 2.11, and a segment of a FORTRAN program which would perform boxcar averaging on an array of N

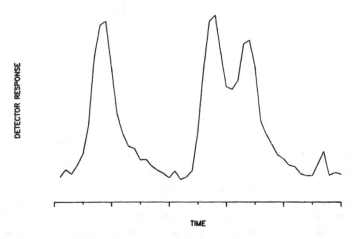

FIG. 2.11.   The effect of boxcar averaging on the analog signal shown in figure 2.7.

points over subsets with M points each is listed in figure 2.12. When boxcar averaging is applied in this way, the original data set is reduced in size. In the above example the 1000 data points are replaced by a set of 100 points.

There is another way to apply boxcar averaging with preservation of the starting number of data points; this method is known as the *moving-window average* (Dulaney, 1975). As with the boxcar technique, a subset of the array is averaged to yield a new point; however, this point does not replace the entire subset but only its central point. Hence the number of data points is not reduced as in the previous example. Subsequent subsets are formed by dropping the first point of the previous subset and adding the next point following the last point in the previous subset. Because each averaged point is calculated from a subset of points consisting of an equal number of points that precede and follow it, the technique cannot be applied to the initial and final points in the set. For example, if the subset size was 11, then the moving-window average could not be applied to the first and last five points in the data set. Often the original data points are substituted in these positions. The moving-average technique has the noise reduction advantages of boxcar averaging without the reduction in effective sampling rate (and hence resolution) of that technique (Dulaney, 1975).

Weighted digital filtering is another technique used to separate noise from quantitative information in data signals. Like the moving-window technique, this procedure uses subsets of the array of data points as windows in which a numerical filter is repeatedly used. The difference is that in weighted digital filtering a set of weighting factors, or coefficients, is used to give different weights to the preceding and succeeding data points around the central point in the window when calculating a new value for the central point. This may be expressed as

$$y_i = \left(\sum_{k=-m}^{m} c_k \cdot y_{i+k} \right)/K \qquad (2.2)$$

where

$$m = (M + 1)/2$$

$$K = \sum_{k=-m}^{m} c_k$$

M is the number of points in the window, and the c_k's are the weighting coefficients. The moving-window average is a special case of weighted digital filtering in which all the c_k's are equal to 1.

The power of weighted digital filtering is in the choice of the weighting coefficients. Given a noisy analog signal, the experimenter would tend to draw a line through the noise that visually best fits the data. Numerically, this can be done by the method of least squares, which has been applied by Savitzky and Golay (1965) to the tabulation of sets of weighting coefficients (c_k's) that can be used for weighted digital filtering. The advantage

```
C        THIS IS A PARTIAL LISTING OF A FORTRAN PROGRAM TO PERFORM
C        BOXCAR AVERAGING ON AN ARRAY 'Y' OF 1000 POINTS WHICH HAVE BEEN
C        PREVIOUSLY ACQUIRED. EACH SECTION OF TWENTY POINTS IS AVERAGED
C        TOGETHER AND SAVED IN A SEPARATE ARRAY 'YP' OF 50 POINTS.
         REAL Y(1000),YP(100)
         DO 200 I=1,1000,20
         X=0.
         DO 100 J=1,20
         X=X+Y(I+J-1)
100      CONTINUE
         YP(I/20+1)=X/20.
200      CONTINUE
```

FIG. 2.12. A partial listing of a FORTRAN program used to filter the analog signal in figure 2.7 by the boxcar averaging technique to obtain the signal shown in figure 2.11.

to using such coefficients is that as much noise as possible is removed without unduly degrading the underlying information. For example, when using a moving-window average (all c_k's = 1) to filter a sharp peak, the peak height would be degraded. A segment of a FORTRAN program that applies an 11-point weighted digital filter to an array of raw data using coefficients determined by Savitzky and Golay (1965) is illustrated in figure 2.13, and the results of applying this filter to the analog signal in figure 2.7 is shown in figure 2.14. Weighted digital filtering is a powerful method for noise reduction in data signals and is widely used in most analytical instrumentation. It works best when the signal is digitized at high densities and

```
C        THIS IS A PARTIAL LISTING OF A FORTRAN PROGRAM TO PERFORM
C        WEIGHT DIGITAL AVERAGING ON AN ARRAY 'Y' OF 1000 POINTS WHICH
C        HAVE BEEN PREVIOUSLY ACQUIRED. A MOVING WIDOW OF 19 POINTS IS
C        EMPLOYED AND THE COEFFICIENTS OF SAVITZKY AND GOLAY (1965) ARE
C        USED. THE SMOOTHED ARRAY IS STORED IN 'YP'.
C
         REAL Y(1000),YP(1000),C(19)
         DATA C/-136.,-51.,24.,89.,144.,189.,224.,249.,264.,269.,
       1 264.,249.,224.,189.,144.,89.,24.,-51.,-136./
         DATA CSUM/2261./
         DATA M/19/
         DO 200 I=M/2+1,N-M/2
         X=0.
         DO 100 J=1,19
100      X=X+C(J)*Y(I-M/2-1+J)
         X=X/CSUM
200      CONTINUE
C        THE FIRST AND LAST 9 POINTS OF Y CAN NOT BE SMOOTHED BY THIS
C        TECHNIQUE, SO THEY ARE TRANSFER DIRECTLY INTO 'YP'
         DO 300 I=1,M/2
         YP(I)=Y(I)
         YP(N-I+1)=Y(N-I+1)
300      CONTINUE
```

FIG. 2.13. A partial listing of a FORTRAN program used to filter the analog signal in figure 2.7 with an 11-point weighted digital filter to obtain the signal shown in figure 2.14.

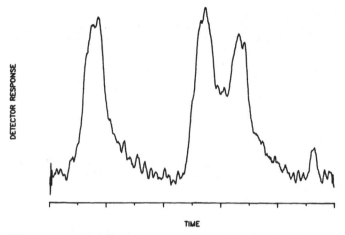

FIG. 2.14.   The effect of an 11-point weighted digital filter on the analog signal in figure 2.7.

the number of points is chosen so that there is no more than one inflection point of the data signal within a window.

Ensemble averaging, like boxcar averaging, is a simple averaging technique, which differs from the previously discussed methods in that it uses several sets or replicates of the data signal to be averaged. That is, the experiment is repeated several times. Based on the assumption that the data sets or arrays are repetitive, they are averaged together point by point. The amplitude and phase of the noise relative to the data in the arrays should be random, so that the noise average should approach zero. The primary requirement for effective use of this technique is that the independent variable, usually time, can be repetitively controlled. For experiments in which computerized instruments are used, this requirement is often trivial. With instruments that lack computerized control synchronization, such as those with manual sample injection into a GC, synchronization of the runs is more difficult.

The last technique to be discussed, *frequency domain analysis,* uses the difference in frequency that usually exists between the data and noise components in an analog signal to identify and filter the noise components. Analog signals are normally acquired at evenly spaced intervals over an independent variable such as time. In order to identify noise components by frequency, the signal is transformed into the frequency domain. Since noise is often present at high and data at low frequencies, elimination of the high-frequency portion of the spectrum, followed by an inverse transformation back to the time domain, smooths much of the noise. The Fourier transform is most often used for the transformations to and from the frequency domain. For digital computers the Fourier transform is normally programmed according to an algorithm published by Cooley and Tukey (1965). The technique has been elaborated by Brigham (1975), and its application to data smoothing on microcomputers has been shown by Anbanel and Oldham (1985). The frequency domain representation $Y(f)$, obtained by the latter's procedure, of the analog signal $y(t)$ (fig. 2.7) is shown in figure 2.15. In this representation the noise appears at the high frequencies (positive and negative) and the data signal at the lower frequencies. The procedure for removing the high frequencies can be represented as a multiplication

$$Y'(f) = Y(f) \cdot F(f) \tag{2.3}$$

where $F(f)$ represents a filter function. The ideal function might seem to be

$$F(f) = 1 \qquad \text{for } f \leqslant f_c \tag{2.4a}$$

and

$$F(f) = 0 \qquad \text{for } f > f_c \tag{2.4b}$$

where f_c represents a cutoff frequency whose value is above the frequency components of the data signal but below the noise components. However, such a sudden cutoff can lead to a false accentuation of frequencies corresponding to transform points in the vicinity of f_c. To avoid this, a more gradually changing filter function is used, for example, a quadratic function such as

$$F(f) = 1 - (f/f_c) \qquad \text{for } f \leqslant f_c \tag{2.5a}$$

and

$$F(f) = 0 \qquad \text{for } f > f_c \tag{2.5b}$$

The result of applying such a function to figure 2.15a, followed by retransformation (inverse Fourier transformation) is shown in figure 2.16. Although a very powerful method, the Fourier transform involves many more calculations than other methods. However, with the continued improvement of microprocessor power at lower costs, the popularity of frequency domain analysis continues to grow. Fourier analysis has been used in many other applications, such as deconvolution of fused GC peaks (Rayborn et al., 1986).

After removal of noise, there are many numerical manipulations that can be performed to obtain useful information from a data signal. For a GC signal these include peak position, height, and area determinations. Although these types of data manipulation may be straightforward programming tasks, the recognition of peaks by a computer program is a nontrivial task, largely because the computer program must be able to discriminate between real peaks and other deviations of the signal from the baseline, such as those due to noise fluctuations (Cooper, 1983). The ability of computer programs to perform these tasks is greatly enhanced by filtering the data signal to improve the S/N ratio.

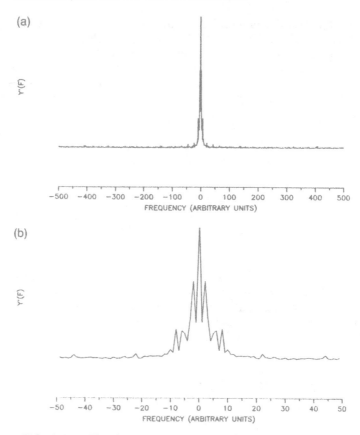

FIG. 2.15.   The frequency spectrum obtained from a Fourier transform of a 512-point digital representation of the analog signal shown in figure 2.7. (a) The full frequency spectrum; (b) an enlargement of the low-frequency portion.

HEAT PENETRATION TESTS: A CASE STUDY

This case study example will illustrate the use of computer-based data acquisition systems for laboratory automation of heat penetration tests in the food canning industry. These tests provide the critical temperature-time data required for the design of safe thermal processes for achieving commercial sterilization of canned foods. Thermal processing consists of heating food containers in pressurized steam retorts at a constant retort temperature for a specified process time. In commercial processing operations, only the retort temperature and process time are control variables that can be adjusted. The degree of bacterial inactivation required for commercial sterility is a function of the internal temperature-time history of the product at the center of the container, which forms the critical data

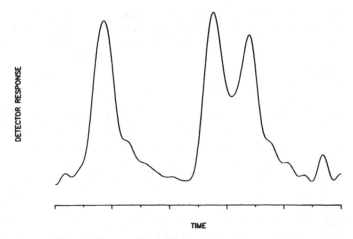

DETECTOR RESPONSE

TIME

FIG. 2.16.   The inverse transformation of the frequency spectrum in figure 2.15 after application of the filter function of equation (2.5) with $f_c = 20$.

base upon which a safe process time can be specified for a given retort temperature.

Traditional Methods

Obtaining such temperature-time data at the can center is the object of laboratory heat penetration tests conducted in pilot-scale retorts for processing test product cans equipped with specially designed thermocouples for this purpose. In traditional laboratories the thermocouple lead wires are passed through a pressure-seal packing gland in the wall of the retort and connected to a recording potentiometer, which converts the milivolt signal from the thermocouples into the corresponding temperature reading and provides a strip chart recording of temperature over elapsed time.

These time-temperature data must be analyzed in order to find the process time that will achieve the desired level of bacterial lethality (F_0) at the can center. Because the lethal rate of bacterial spores is temperature-dependent and the center temperature varies over time, the final lethality can be determined by integrating the lethal rate at short time intervals over the process time used in a given heat penetration test. This is known as the *general method* of thermal process determination from heat penetration data analysis. Additional heat penetration tests using longer or shorter process times can then be run until the precise process time that produces the required lethality is found. Because of the large number of repeated calculations, this method of data analysis is tedious and time-consuming. The calculation of F_0 by integration can be quickly performed by computer, but the attention to detail that is required when inputting the raw data prior to each program execution is also time-consuming and error-prone.

A more popular traditional method for heat penetration data analysis is the use of *Ball's formula* for calculating thermal process time at a given retort temperature. The formula is derived from the mathematical equation of the straight-line portion of the temperature-time relationship at the can center when plotted on inverted semilogarithmic graph paper. The ordinate axis is labeled 1° below retort temperature at the top, with each logarithmic cycle representing a tenfold decrease in temperature when proceeding downward. Since the center temperature can never theoretically reach the retort temperature, the curve will extend as a straight line for any length of heating time beyond the initial response lag. Parameters related to the slope f_h and intercept j of this straight line are required in order to use the formula, along with a set of tables forming an integral part of the method. Although this formula method relieves some of the tedium experienced in the general method, it still requires careful transformation and plotting of the raw data onto graph paper in order to complete the necessary data analysis for determining the process time from heat penetration tests.

Computer-Automated Methods

From the information presented earlier in this chapter on data acquisition in the laboratory, it is readily apparent that considerable advantages in time-saving efficiency can be gained by use of a computer-based data acquisition system for conducting heat penetration tests. A simple system for this purpose is illustrated schematically in figure 2.17. Thermocouple lead wires from the retort are connected to a data logger or similar device capable of converting analog thermocouple signals into digital signals, which can be read by a computer as electronic input data. Data loggers are commercially available instruments, which can serve as stand-alone strip chart recorders but also contain A/D circuit boards for the added flexibility of feeding data directly to a computer for automated data analysis.

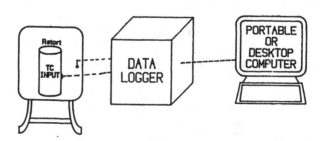

FIG. 2.17.   Schematic illustration of automated data acquisition system for heat penetration tests on canned foods.

With a system of this type, signals sent by thermocouples measuring retort temperature and temperature at the center of the test product are fed directly into the computer, completely bypassing the need for an operator to input the data at a terminal keyboard and eliminating the possibility of operator error in data input. Once fed into the computer, the data can be stored in memory and called up on demand for direct printout in tabular or graphical form. The data can also be further processed through various transformations and calculations in accordance with preprogrammed software to produce output at any intermediate stage of data analysis desired.

As an example, figure 2.18 is a direct computer printout of heat penetration data that had been programmed to be plotted on an inverse semilogarithmic graph in order to obtain the parameters f_h and j (the slope and intercept, respectively, of the straight-line portion of the curve) for subsequent use in calculating process time by the Ball formula. The software package for this particular application offers the added feature of operator control in positioning the straight line on the screen before the computer is instructed to compute the required parameters. The computer can be

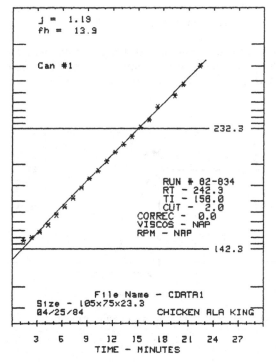

FIG. 2.18. Computer printout of electronically fed heat penetration test data plotted on an inverse semilogarithmic graph. Computed values are shown for parameters f_h and j needed to calculate process time by the Ball formula for a given product test. (Courtesy Central Analytical Laboratories, Inc., Kenner, LA).

instructed to generate the line that statistically best fits the data points, but alternatively, the operator can move the line on the screen so that it lies conservatively below all the data points, as many food scientists prefer to do for an added safety factor.

Figure 2.19 is a computer printout of heat penetration data handled in a different way. With this software package both retort temperature and can center temperature are plotted against time without any data transformation. This allows the operator to see what has actually happened throughout the process test. As the data are being read by the computer, additional software instructions call for calculation of lethality F_0 by calculating the lethal rate at the measured temperature over each time interval between temperature readings. As a result, the accumulated F_0 is known at any time during the process and can be plotted on the graph along with the temperature histories to show the value reached at the end of the process. In this way another test can be quickly run for a longer or shorter process time to determine the corresponding higher or lower F_0 achieved. By examining the results from both tests, the desired process time for the target F_0 value can be closely estimated and then quickly tested. The results of two such heat penetration tests are shown superimposed on each other in figure 2.20. These results show that the first test, with a process time of 68 minutes, produced an F_0 value of 6, while the second test, with a process time of 80 minutes, produced an F_0 value of 8; this result suggests that a target F_0 value of 7 will likely be achieved by a process time in the order of 74 minutes. This conclusion could then be confirmed by running a test at the suggested process time and examining the resulting F_0 value.

In this chapter the benefits of computer-based data acquisition systems in the laboratory have been illustrated by case study examples involving gas chromatography for food analysis and heat penetration tests for design

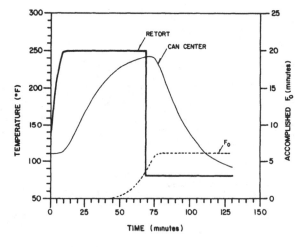

FIG. 2.19. Computer plot of electronically fed heat penetration test data showing both measured retort temperature and product can center temperature as a function of time, along with the calculated product lethality (F_0) accumulated at the can center over time for instant process evaluation.

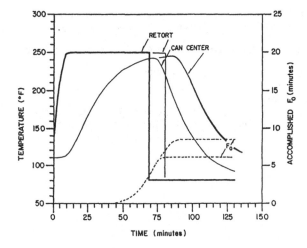

FIG. 2.20. Computer plots of electronically fed heat penetration data from tests using two different process times superimposed on each other for estimation of process time needed to achieve a specified F_0 value.

of thermal processes. These examples represent two widely different types of data, sensors, and methods of data analysis. Countless other examples of laboratory applications can be found throughout the food industry. Hopefully, the information presented in this chapter will inspire the recognition of application opportunities not yet identified and a basis upon which to proceed for further development of such opportunities.

Subsequent chapters will deal with computer applications in production control, control of unit operations, process modeling or simulation, and process optimization. Each of these chapters will include case study examples related to various food processing operations or food industry applications. In order to provide a continuous thread throughout these chapters, reference to thermal processing operations in food canning will appear as a common case study example in each chapter. By focusing on these thermal processing–related case studies in each chapter, it should be possible to see how the increasing use of computers at all stages from the laboratory to the production plant can lead to significant improvements in production efficiency and quality assurance for a specific subsector (canned foods) of the food industry.

REFERENCES

ANBANEL, E. E., and OLDHAM, K. B. 1985. Fourier smoothing. Byte 10(2):207–218.

BRIGHAM, E. O. 1975. The Fast Fourier Transform. Prentice-Hall, Englewood Cliffs, NJ.

COOLEY, J. W., and TUKEY, J. W. 1965. An algorithm for the machine calculation of complex Fourier series. Mathematics of Computation 19(90):297–301.

COOPER, J. W. 1983. The Minicomputer in the Laboratory. John Wiley & Sons, Inc., New York.

DESSY, R. E.   1986.   The PC connection. Part II. Analytical Chemistry 58:793A.

DULANEY, G.   1975.   Computerized signal processing. Analytical Chemistry 47:25A–32A.

EWING, G. W.   1985.   Instrumental Methods of Chemical Analysis. McGraw-Hill, New York.

RAYBORN, G. H., WOOD, G. M., UPCHURCH, B. T., and HOWARD, S. J. 1986.   Resolution of fused gas chromatographic peaks by deconvolution with extension of the Fourier spectrum. American Laboratory 18(10):56–64.

SAVITZKY, A., and GOLAY, M. J. E.   1964.   Smoothing and differentiation of data by simplified least squares procedures. Analytical Chemistry 38:1627–1639.

3

Computer Control in the Food Processing Plant

This chapter focuses on taking the computer into the processing plant for the control of processing operations on the plant floor.

REVIEW OF PROCESS CONTROL SYSTEMS

Traditional Relay Systems and Automatic Controllers

In order to appreciate the role of computers in process control, it is necessary to have some basic familiarity with the more traditional process control technology that has been in place throughout much of the food industry. Two basic components are required for automated process control: a means by which important process conditions, such as temperature, pressure, and flow rate, can be held constant at a predetermined set point; and a means by which the sequence of operation can be controlled. To illustrate the second requirement, one must be certain that ingredient A has been added and mixed before ingredient B is admitted to a blend tank; one must also be sure that a storage tank is empty and clean before a valve that would admit a finished product can be opened (otherwise the product could be inadvertently admitted to a full tank).

In traditional process control the requirement of constant process conditions is met by use of automatic controllers, which operate either pneumatically or electronically, as shown in figures 3.1 and 3.2, and have been the backbone of most industrial process control systems for many years. These controllers form an integral part of a *control loop,* in which they receive a signal from a sensor measuring the parameter to be controlled (such as temperature or pressure) and transmit a signal to a control valve, causing it to open or close in accordance with the corrective action required to minimize the error signal being received from the sensor. The rate at which a controller responds to the error signal is a function of the proportional, integral, and derivative (PID) components of the system response time relative to preset adjustments for these functions in the controllers themselves. Engineers can use process control theory to analyze the system

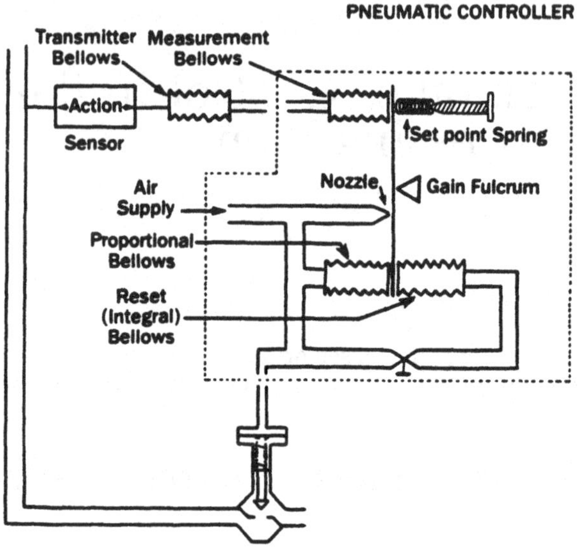

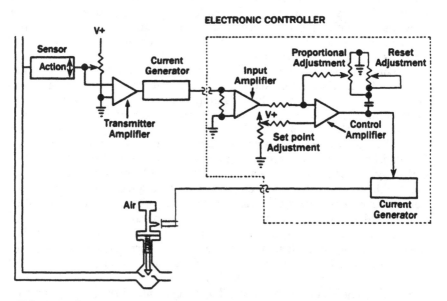

PNEUMATIC CONTROLLER

Transmitter Measurement
Bellows   Bellows

Action

Sensor

Set point Spring

Air
Supply

Nozzle   Gain Fulcrum

Proportional
Bellows

Reset
(Integral)
Bellows

FIG. 3.1.   Operation of
a pneumatic controller.
(From Shaw and Mc-
Menamin, 1984.)

ELECTRONIC CONTROLLER

V+

Sensor

Action

Current
Generator

Input
Amplifier

Proportional
Adjustment

Reset
Adjustment

Transmitter
Amplifier

V+

Set point
Adjustment

Control
Amplifier

Air

Current
Generator

FIG. 3.2.   Operation of an electronic controller. (From Shaw and McMenamin, 1984.)

response curve mathematically and determine the optimum setting for these adjustments, which are referred to as *gain, proportion,* and *reset* on the controllers. These adjustments are actually accomplished by changing the tensions on various bellows in the pneumatic controllers or the electrical resistance on variable resistors in the electronic controllers.

The second feature of process control is the need to achieve the proper sequential logic among all the control loops in the system. This is traditionally accomplished through the use of hard-wire relay circuits governed by timers and counters, which automatically turn various circuits on and off at predetermined times or events. Process control engineers know how to design these circuits through the use of *ladder diagrams,* which show how wires need to be connected between the various gauges, switches, valves, and controllers on the plant floor and the various timers and counters on the main control panel. A familiar example of this type of control system is found on major household appliances such as clothes washers or dishwashers which automatically carry out a sequence of operations when the control dial is set to a selected cycle of operation.

The major advantage of these hard-wire relay systems is that they enjoy a long history of proven reliability and are well understood by experienced process engineers in the food industry. Their major disadvantage is their limited flexibility. Essentially, once a process is designed and installed, the hard-wire relay control system is close to being "chiseled in stone." Often, a major plant shutdown needs to be scheduled in order to accomplish the necessary rewiring and circuit connections required to add a new unit operation or to change the sequence of operations in the process. This often inhibits the timely implementation of process improvements and efficiency recommended on the basis of the company's research and development efforts.

Computer-Based Process Control Systems

From what we already know about computers and how they can be interfaced with instrument sensors and devices through I/O process modules with A/D or D/A converters, it is not difficult to envision systems in which computers are programmed to read data from sensors and send signals to process control devices on the plant floor. In fact, this capability lies at the heart of all computer-based process control systems. The general configuration of such a system is illustrated schematically in figure 3.3, which also illustrates how the same computers could be connected to various memory storage devices and other communication modules for data storage and retrieval, operator interface, and communication with other computers and controllers throughout the plant. Perhaps the feature to be most appreciated is the fact that, once all the bus cables have been connected, process changes can be easily made by simply reprogramming the computer through keyboard instructions.

In spite of these perceived advantages, computer-based process control

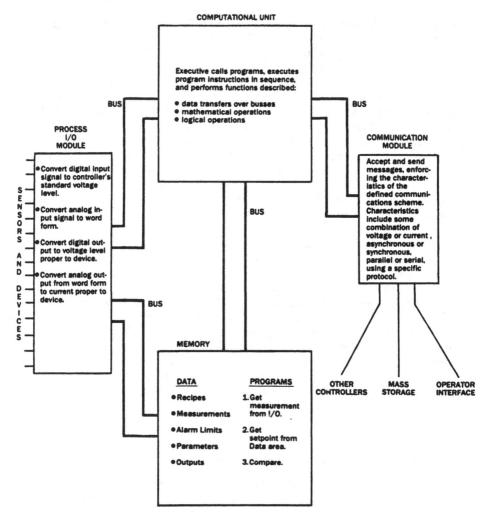

FIG. 3.3.   General configuration of a computer-based process control system.

systems were slow in making any major inroads in the food processing industry. Computers were little understood by most production plant engineers. A natural fear of the unknown, coupled with the generally conservative nature of the food industry, tended to create an attitude of "let's wait until we see someone else do it first." Perhaps the single most important technical development that has led to the increasing acceptability of computer-based process control systems has been the introduction of the microprocessor-based programmable logic controller (PLC), also called, more simply, programmable controllers.

A PLC is a controller that can perform the same logic, timing, and counting functions as relays, with the same simplicity of operation and ability to withstand harsh plant conditions but with vastly greater flexibility. It is designed to accept programs written in a programming language similar to the ladder logic that electricians use to describe relay circuits and is actually a digital computer that no one would think of as a computer.

Programmable controllers have steadily grown in sophistication and capabilities, benefiting from advances in computer technology while maintaining their image as "noncomputer" controllers. They began as flexible relay replacements, good for repetitive manufacturing applications, but gradually developed the capabilities needed for process control. With use of microprocessors, their functions expanded from strictly digital operations (interlocking, timing, counting) to arithmetic, data transfer, matrix, message, and packaged PID algorithm functions, all within the ladder logic framework. Larger memories meant room for recipe storage, communication software, and more I/O. Their designers have even recognized the limitations of the ladder logic programming language and added capabilities for higher-level languages.

Therefore, since programmable controllers offered many advantages, were easy to use, were cost-competitive, and were not advertised as computers, they became readily accepted as modern improvements to traditional relays and automatic controllers in industrial process control applications. Once they were in place and operational, their actual benefits often exceeded initial expectations, and process control engineers could begin to appreciate the further advantages of linking programmable controllers to higher-order computers and to each other for more sophisticated control system configurations.

CONTROL SYSTEM CONFIGURATIONS

The various combinations and permutations in which computers and/or programmable controllers can be linked together in forming a computer-based process control system fall into three basic system configuration types, commonly referred to as *dedicated, centralized,* and *distributed* control systems. A simple but hypothetical food process described by Hyde and Clem (1985) was useful in schematically illustrating how these control system configurations differ from each other. A similar approach will be used in the following subsections, where the relative advantages and disadvantages of each configuration will also be discussed briefly.

Dedicated Control Systems

A dedicated digital control system is one that is based strictly on the use of individual microprocessor-based control units (programmable controllers), each dedicated to the control of a specific unit operation, such as the continuous blending of several ingredients into a formulated batch or controlling the temperature in a heat exchanger for a heat treatment process. These digital controllers have simply been installed to replace traditional analog controllers, timers, and counters in the dedicated control of each unit operation. They do not communicate with each other or with any other computer but instead simply receive on/off instructions from a traditional central control panel through existing hard-wire relay circuits.

Figure 3.4 illustrates how such a dedicated digital control system would be applied to a hypothetical food process made up of four unit operations, namely raw material storage, batching, heat treatment, and packaging. Each unit operation is controlled by a dedicated programmable controller or similar microprocessor-based control unit, which in turn is controlled directly from a main control panel through traditional relay circuit logic. As such, this type of system offers minimum disruption when retrofitting an existing relay-based control system and often provides a comfortable entry-level experience with computer-based control systems for most companies. This system would have a low to moderate degree of sophistication depending upon the capabilities of the individual controllers and the supervisory control system. Dedicated digital controllers often have only the ability to supply other devices with discrete output signals and receive discrete input signals; thus data logging, report generation, and automatic setpoint adjustment capabilities are typically limited. Also, where the controllers have the ability to communicate on a higher level with other devices, a common communications language and protocol for the entire system may be very difficult to obtain.

The operator interface for this system, excluding the main control panel, would probably be relatively rudimentary and inflexible. An interface would likely be present at each dedicated digital controller for use in the alteration of process variables and the like. Because of this, the design of the operator interface would be, to some degree, dependent on the individual controller's interfacing capability.

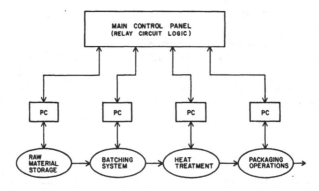

FIG. 3.4.   Dedicated control system configuration (PC = programmable controller).

Dedicated digital controllers are widely used and have remained popular since their introduction to the food industry for several reasons. First, they can be added to an existing process without requiring much, if any, alteration to the rest of the control system. Additionally, dedicated digital controllers can typically be programmed in a "fill-in-the-forms" manner, where the end user is only required to enter setpoints and other recipe data. This is both a strength and weakness of this type of device. Because of their ease of programming the devices are often not very flexible. That is, this type of controller is usually designed for one particular type of equipment or process, and thus its operating sequence is preset and unalterable by the end user.

Centralized Control Systems

A centralized control system is the type of control system that usually came to mind when computer control was first proposed for industrial processing operations. In its simplest form this system consisted of a single mainframe computer or large minicomputer located in a central control room. It was in direct control of all the various processing operations by being linked to remote I/O modules located close to processing equipment on the plant floor. The I/O modules contained all the ADCs necessary to convert analog signals received from instruments and sensors on their assigned process equipment into digital signals interpreted by the computer. Using the appropriate control logic programmed into the computer for each respective process operation, the computer would communicate some control decision back to the appropriate I/O module, which would in turn convert the signal into the form necessary to actuate the controlling devices on the process equipment.

Figure 3.5 illustrates how such a centralized control system would be applied to the hypothetical food process described earlier; it also suggests the use of modern programmable controllers in place of I/O modules, as

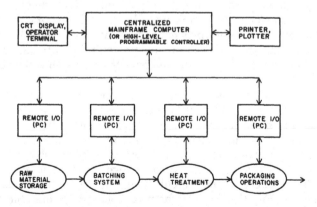

FIG. 3.5. Centralized control system configuration.

well as the use of a high-level centralized digital controller in place of the mainframe or minicomputer for many applications. A historical disadvantage to centralized systems was the constant fear that, in the event of any single failure in the central computer, a total plant shutdown would occur unless an expensive standby computer of equal capacity were ready to take over. The expanding capabilities of sophisticated programmable controllers now make it possible for considerable "sharing of the load" between supervisory and subordinate controllers in a centralized system and have come a long way in minimizing the earlier cost disadvantages of this type of system. The system could have a moderate to high degree of sophistication depending upon the capabilities of the controller and the peripherals that it supports. For instance, the controller, depending upon its level of sophistication, can support such peripheral devices as video displays, printers, data loggers, color graphics terminals, and a mini- or microcomputer interface.

The operator interface for this system can be very flexible and range from a moderate to high degree of sophistication. Because the controller can support a wide range of interfacing peripherals and remote I/O modules, the system can be designed to include substantial capabilities in the areas of report generation, data logging, and operator interface configuration and layout.

Direct centralized digital controllers have many strengths, including costs that are very competitive with those of other types of systems; they can be less expensive than other systems with less capability. The cost savings are due in part to reduced field and internal panel wiring requirements and typically to relatively moderate software costs. This is because programmable controllers can often be programmed in the relay logic language that resembles the relay wiring ladder diagrams familiar to plant engineers and maintenance personnel, as explained earlier. Also, the documentation of the software for these types of controllers can be very thorough.

The flexibility of a direct centralized digital control scheme is also a major advantage. Systems of this type can typically cope with changes in the configuration of the process merely by software revisions. Also, additional logic and I/O capacity are usually not very expensive, and added process capacity may sometimes not require any new controller hardware. Finally, programmable controllers have proven reliability, and redundancy in the form of a permanently installed backup on-line controller is available with some units.

The major pitfall of this scheme is that if the centralized controller fails (without redundancy), the facility will come to a complete stop. Another problem is that most programmable controllers have limited memory capacity, and the available memory in one or a few controllers might be insufficient for a given application. Third, programmable controllers typically are not easily programmed to perform arithmetic operations, and thus any "number crunching" calculations can be very laborious and memory-consuming. Finally, the interfacing of some programmable controllers to higher-level devices is not very easy, and seemingly simple tasks such as data transfer and reporting can be impractically difficult.

Distributed Control Systems

Distributed digital control schemes consist of a group of controllers (which may be programmable controllers or computer-based remote I/O units) that can be programmed with a great degree of flexibility. These controllers can "talk" to each other via a communication network. Each controller is often assigned to a functional area so that it can be relatively independent of the other controllers in the system. Communication via the communication network is only required in instances such as process-to-process interlocking and process data up/down loading to or from the host computer. A schematic representation of distributed digital control as applied to the example food industry process is depicted in figure 3.6. Each of the functional areas of the process is served by an individual controller, and these controllers, in turn, talk to each other and to a host computer via a communication network. Each of the controllers and its I/O instrumentation can be located in close proximity to the equipment that it controls, which thus saves cost by reducing the amount of field wiring required. This system could have a high degree of sophistication as compared with the two other schemes discussed earlier. Its ability to support the intercommunication of different types and levels of devices is a major strength. Such peripherals as printers, plotters, color graphics, interactive operator consoles, hard-disk-drive mass storage units, and interfaces to higher- and lower-level devices are often offered as standard equipment with this type of system.

The operator interface for these systems would likely consist of an interactive operator's console, which would combine graphics with control

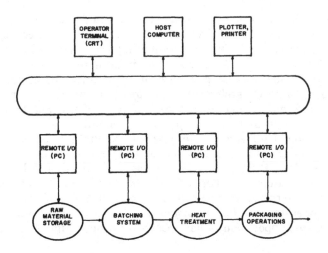

FIG. 3.6.   Distributed control system configuration.

functions. This interface can be centrally located, with small display/control panels located near each of the unit operations. Distributed digital control schemes have not been widely applied until recently owing to several factors, including the relatively high hardware costs as compared with the other two schemes. Additionally, the software costs for distributed systems can be high because of all the interaction between devices. The nature of the programming of distributed systems can vary, but usually the software can be developed in a high-level language, some of which is specific to the hardware itself. The documentation and development of the process control and system software can be quite sophisticated, as some systems offer a text editor–based development system containing functions typically available in word processing packages. Since a host computer is integrated into the system, functions such as optimization, modeling, and other tasks that involve a large amount of "number crunching" are usually readily available. Distributed digital control schemes can be very flexible with regard to both initial configuration and subsequent process changes. As with the centralized scheme, these changes can sometimes be dealt with through software revisions alone.

Integrated Control Systems

When distributed control systems are in place in various processing sections of a food production plant, they can often be integrated into a larger management information system, which can also include administrative, production, marketing, quality control, and maintenance functions of the company's business operations. In integrated systems host controller and subcontroller roles do not apply. Instead, full process controllers, sometimes called *process control modules* or PCMs, independently perform continuous, sequential, and communications functions with mainframe computing power.

PCMs can act as stand-alone controllers for multiple process units. An auxiliary minicomputer (AUX) can perform auxiliary functions such as configuring data bases, editing and compiling programs, loading data bases and programs to the PCM, managing peripheral devices such as operator consoles and disks, executing supervisory programs, and communicating with other information systems in the company's business operations. Since the AUX performs no direct process control, it poses no threat to the process if it fails, and a redundant PCM provides economical backup.

If processing requirements are broader, the PCM's communication capability can be used to create a system of independent but unified multiunit controllers. If any PCM fails, the others can continue their full control and communication functions. The minicomputer provides the same auxiliary services with no threat to control or to communications if it fails. To achieve extensive operator interfacing, *display control modules* (DCMs) can be put in the link to support additional operator consoles. Figure 3.7 shows the

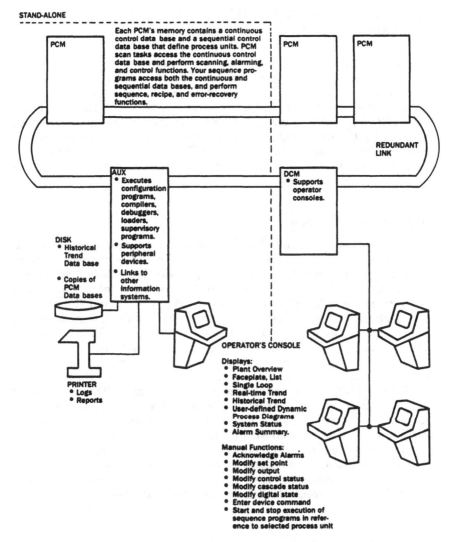

STAND-ALONE

PCM

Each PCM's memory contains a continuous control data base and a sequential control data base that define process units. PCM scan tasks access the continuous control data base and perform scanning, alarming, and control functions. Your sequence programs access both the continuous and sequential data bases, and perform sequence, recipe, and error-recovery functions.

PCM

PCM

REDUNDANT LINK

AUX
• Executes configuration programs, compilers, debuggers, loaders, supervisory programs.
• Supports peripheral devices.
• Links to other information systems.

DCM
• Supports operator consoles.

DISK
• Historical Trend Data base
• Copies of PCM Data bases

PRINTER
• Logs
• Reports

OPERATOR'S CONSOLE

Displays:
• Plant Overview
• Faceplate, List
• Single Loop
• Real-time Trend
• Historical Trend
• User-defined Dynamic Process Diagrams
• System Status
• Alarm Summary.

Manual Functions:
• Acknowledge Alarms
• Modify set point
• Modify output
• Modify control status
• Modify cascade status
• Modify digital state
• Enter device command
• Start and stop execution of sequence programs in reference to selected process unit

FIG. 3.7.   Fully integrated control system. (Courtesy EMC Controls, Inc.)

system. Process control schemes can take many forms and configurations, but the most important consideration is that of matching the facility and process control requirements with the capabilities of the control scheme. Each of the control schemes reviewed above is popular to some degree in the food industry today. However, it appears that the control schemes of the future will continue to move toward more sophistication in the form of total facility integration.

BENEFITS OF COMPUTER-BASED CONTROL SYSTEMS

One of the most interesting results often documented in reports of companies that have installed computer-based process control systems is that the benefits actually realized frequently exceed the benefits that were initially anticipated or identified as a basis for project justification. This is because it is difficult to identify a priori all the various ways in which computer control can bring about improvements in processing operations. Shaw and McMenamin (1984) identified five general areas in which significant benefits from computer control are often realized. These are discussed in the following paragraphs.

Increased Production

There are many ways in which a computer can improve production. If many products are being manufactured and shared equipment must be scheduled, the computer can optimize the production schedule and equipment assignments to achieve the greatest product throughput and equipment utilization. The computer can even take into account the ordering of production batches so that vessel cleaning is minimized. If the computer controls the charging of multiple reactants, it can often cause them to be charged simultaneously rather than sequentially and thus cut down the charge time. Computer control can provide a time-out on operator choices in noncritical stages and in many cases the computer can check the permissive conditions so that no operator interaction is required at all. The computer can also track operator response times and provide added levels of prompting if excessive operator delays are detected, and it can automate various manual operations and reduce overall batching time. If the plant has a single additive system, it is quite common for only one vessel to be allowed to charge at a time. If the additive distribution system does not prevent it, a computer can often allow simultaneous vessel charging because the computer is fast enough to track and control multiple charging operations.

Improved Consistency

Computers are very good at doing the same thing over and over again in exactly the same way. This is a very important consideration as minor variations in processing can cause a product to be off specifications and worth less than it should be. In charging a vessel to a defined recipe, the computer is consistent from batch to batch and never makes an error (provided that the additive system is under computer control and is mechanically accurate enough to provide consistent results in response to consistent control commands). The computer's ability to operate in a dual mode allows control loops to be very accurately tuned so that product quality can be personnel for performance consistency and warn when valves start sticking or slowing down, when sensors fail or drift, and when a given operational shift has more off-specification production and downtime than others.

Improved Efficiency

By using the computer to control heating and cooling loops, it is often possible to operate at maximum values of reaction conditions. This minimizes both the batching time and the cooling/heating energy requirements (pump energy usage, throttle losses from valves, heat exchanger fan energy, etc.). Computer control of reactants is often more accurate than noncomputer control. An excess of reactants is often nonrecoverable, and its presence may require excessive quantities of reaction inhibitors.

Safety

Throughout its normal sequencing of a process, the computer can make safety checks that verify the correctness of operations. If a valve sticks open and too much of a raw material is charged, if a temperature starts rising too fast, or if a level does not rise once a fill valve is opened, the computer can initiate fail-safe procedures automatically. In addition, the computer can record operator actions and computer actions, so that in the event of a failure the steps leading to the failure can be analyzed and the failure avoided in the future. The computer can often be used to "double check" direct measurements by computing an expected result from a related measurement. For example, the flow into a vessel can be checked by the level changes in the vessel, and a temperature rise can be seen in a pressure jump. One key area in which safety can be improved is the verification of operator input reasonableness. For example, the operator who dials in a setting of 1000 pounds when 100.0 pounds is needed should be told to try again by the process control computer.

As for computers themselves, computer systems do fail. People have been understandably gun-shy after some of the early computerized process control experiments. However, computer hardware has become fairly dependable (and cheap), and the use of redundant hardware and distributed architectures has made computer-based control systems very reliable.

Cost

The cost of a computer system must be weighed against its benefits, especially for an existing plant, since the advantages of a computer system may not provide improvement over current control capabilities.

In a new facility a computer system is often less expensive than conventional instrumentation. It is a myth that the cost of a computer is not competitive with that of conventional analog controls, although on a loop-by-loop basis computers have not yet achieved the cost level of analog controllers. However, when data logging, sequential control, and the cost of making an elaborate process simulation panel with annunciators are added in, computers begin to look very attractive. When the factors of increased production, safety, consistency, and efficiency are also considered, the computer system usually wins hands down. While the long-term picture is better with the computer, the initial cash outlay to install a computer system often exceeds the cost of conventional controls. Unfortunately, the long term is not always considered when a purchase decision is being made.

SYSTEM SPECIFICATION AND VENDOR SELECTION

A necessary first step to suggesting guidelines that should be followed once a decision has been made to purchase and install a computer-based process control system is to prepare a detailed specification that describes what such a system is required to do. This specification can be sent out to system vendors for bids or used as a blueprint for a large company's own internal engineering group.

One of the most important considerations in preparing a specification is to involve the correct people in its creation. Every group whose work will be affected by the system should be given a chance to comment on the specification and should be prepared to give required support. The following personnel should normally be included in the team that prepares a specification.

1. The *process engineer* must have a thorough knowledge of the food processes involved and their critical process parameters and control problems. This engineer must ensure that the selected system can control the process and should help define the required control schemes.
2. The *equipment engineer* must be totally familiar with the plant equipment and instrumentation and with its maintenance. This person should be empowered to make equipment changes as required for the new control system.
3. The *operations supervisor* has the responsibility for hiring and training plant operating personnel and for the actual operation of the plant and will be responsible for operating the control system.
4. The *applications engineer* is responsible for implementing the control schemes (continuous and sequential) on the selected computer system and for creating user's guides and application documentation. This person is also responsible for creating any special logs and reports required by the plant, and should have a good background in both process control and computers.

The exact titles are not important—only the knowledge and experience that these people bring to the table. As with any project, it is important to have a leader. A good approach is to assign a senior project engineer to supervise and coordinate team activities. The main reason for having such a diverse collection of talents is to ensure that the specification covers all the areas that concern the plant and the capabilities of the system being purchased. The other reason for using all these people is to ensure that the departments they represent all get a chance to provide inputs into the system design.

Some major categories of information, as identified by Shaw and McMenamin (1984), that should be included in a specification are listed below. This list is not to be construed as all-encompassing, but rather as a general guide. In some instances, some of the items may be excluded or the vendor may be instructed to provide a recommended standard.

1. *Process description:* The vendor can best size the system and its needs if a detailed explanation of the process steps and the plant equipment and operations can be provided. Process and instrumentation (P&I) diagrams and mechanical drawings are useful documents for this purpose.

2. *I/O definition:* A detailed list, by equipment area, of all process signals and interfaces, including all unusual signal types and special "smart" devices, should be provided. Spare capacity should be enumerated and described, since there are often questions about spare capacity and actual spare I/O points. Requirements for special hardware or empty slots or racks should be stated. To overlook transmitter powering and signal fusing can also prove expensive.

3. *Cabinets:* Requirements for space or height limitations or special access restrictions should be clearly defined.

4. *Environment:* If the system is to be placed in a hostile environment, the environmental extremes should be specified. This should include descriptions of hostile storage environments prior to installation.

5. *Power:* Computers require clean, regulated power. If it cannot be provided by the user and the vendor must provide it, this should be made clear. Any grounding problems should be spelled out.

6. *Training:* The user should specify what training will be needed and how many people are to be trained; if the user is uncertain, the vendor should be asked for recommendations.

7. *Testing:* Most vendors provide some type of factory acceptance test. If special testing at the factory and/or the plant site is required, the testing to be performed should be described. Most vendors have a standard test procedure, which will often meet most requirements and which the user should ask to review. Process simulation testing, if required, should be thoroughly detailed.

8. *Operator interfaces:* Since the system will work only if the operations people accept it and use it, the operator interface needs should be well defined. All logs, reports, and custom displays should be specified in great detail. If possible, operators should get hands-on experience with all the prescreened vendors' systems. This will help operators

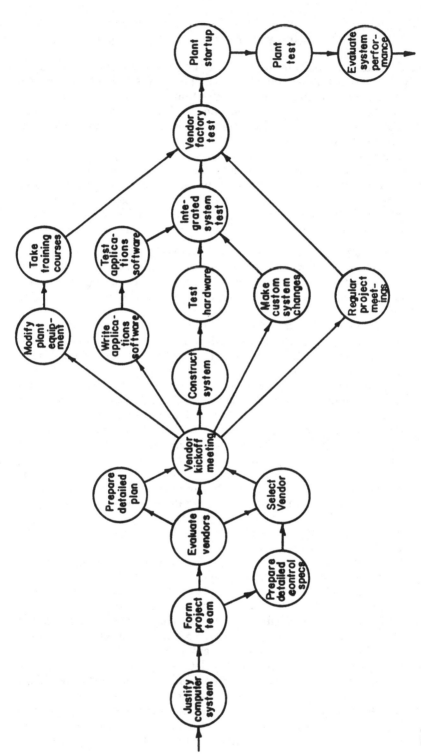

FIG. 3.8.   PERT diagram showing implementation plan for installing computer-based control system. (From Shaw and McMenamin, 1984.)

to develop an importance factor for the available display types and help to specify the displays that are considered essential.

9. *Application assistance:* There should be a detailed description of the application effort involved. Some customers assign their application engineers to vendor facilities during the project so that the engineers become very familiar with the system and the vendor. Quoting application assistance is very difficult unless detailed information can be provided. This should include P&I diagrams, ladder diagrams, flowcharts, recipe information, and other details.

10. *Field assistance:* During the project there will be a need for meetings at the processing facilities, supervision of system installation, start-up assistance, warranty support, plant acceptance tests, and other field assistance. All these activities should be called out and defined so that vendors can include these services in their quote.

The success of most computer-controlled installations has hinged on the abilities of the project team and the quality of their implementation plan. Figure 3.8 provides a rough plan for most computer control projects in the form of a PERT (program evaluation and review technique) diagram for project implementation.

INDUSTRY CASE STUDIES

The five case studies described below illustrate commercial applications of computer-based process control in various sectors of the food processing industry. The examples differ not only in the type of processing operations being controlled but also in their focus on the various aspects of computer control presented thus far in this chapter. For example, the first case study, which deals with vegetable oil refining, focuses on the actual benefits realized from a commercial system. The second study, involving dairy process control, focuses on the technical requirements for specifying a computer control system. The third and fourth examples, which involve distilled liquor and orange juice, respectively, are intended to illustrate the use of computer controls in automated storage and retrieval systems for warehousing of packaged goods and tank farm storage of bulk liquids. The final example, dealing with control of canned food sterilization in batch retorts, focuses on the hardware and software requirements for a centralized computer-based control system.

Vegetable Oil Refinery

The purpose of this case study example is to examine A. E. Staley Manufacturing Co.'s reported experience with computer-based process control of vegetable oil processing in its Des Moines, Iowa soybean oil refinery. The material is borrowed from a presentation by N. J. Smallwood (1985) and reflects the perspective of a former plant manager who was responsible and accountable for the hard measure profit and loss results from the processing operation. The case study will focus on the benefits and shortcomings actually experienced, along with an outline of the strategy that was used to achieve successful results.

Benefits. As discussed earlier in this chapter, there are many potential benefits from the application of computer-based process control but the magnitude of potential improvements is application-specific and will vary by plant. The specific advantages of computer control that were realized by A. E. Staley in its vegetable oil processing operations are outlined and briefly described below.

1. *Material usage:* Precise programmed reaction conditions (temperature, pressure, catalyst addition, and hydrogen gas input) for hydrogenation result in more uniform physical characteristics of each base stock produced. The consequences of improved base stock uniformity are more predictable product blending (mixing) results, less correction requirements for blends (mixes), and a possible reduction in base stock inventory. With conventional control, Staley often maintained a larger base stock inventory in an effort to reduce the variation between batches of the same base stock. More precise, responsive control resulted in minimum usage of materials such as caustic soda, sulfuric acid, bleaching earth, filter aid, catalyst, hydrogen gas, nitrogen gas, and other processing reagents and aids. The hydrogen gas plant operating rate could be programmed according to hydrogen demand, and hydrogen venting could be consistently avoided.

2. *Operating mistakes:* Once the process control system was reliably programmed, mistakes were essentially limited to equipment failures. Consequently, a significant reduction in pumping errors and spills was achieved, and hence the expense associated with the resulting additional inventory, reprocessing, degrading, loss, handling, and cleanup could be substantially reduced.

3. *Diagnosis and troubleshooting:* The capability of trending, logging, and alarming offered in computer process control is invaluable for process diagnostics and troubleshooting. Problems can be quickly identified and resolved.

4. *Yield and quality performance:* Tighter and more consistent process control (a reduction in both under- and overprocessing) obviously provides the opportunity for improved yield and product quality performance.

5. *Energy reduction:* Numerous opportunities are available for energy reduction, including improvements in combustion efficiency and load management. Load management involves the elements of monitoring, trending, alarming, scheduling, and controlling energy loads (steam, electricity, etc.) to minimize use and avoid excessively high peak demands.

6. *Process safety:* By means of computer programming, better process safety can be provided in terms of alarms, interlocking, fail-safe modes, and emergency shutdowns.

7. *Inventory control and reduction:* Continuous tank level reporting and inventory calculation provide an effective data base for both operation and management functions. With close-coupled processes and controlled processing rates to keep the total system in balance, in-process oil inventory can be significantly reduced.

8. *Tank car and tank truck utilization:* A common problem with manual control of tank car and tank truck loading is underfilling. To avoid spillage, operators tend to be conservative and underfill. Computer control combined with on-scale loading allows either precise-weight or maximum filling.

9. *Process capacity:* With computer-linked and -paced operations, design capacity can be consistently achieved.

10. *Performance and accounting reports and operating logs:* The system is capable of providing any type of performance and accounting report and operating log within the scope of the information sources (devices) connected and the programming provided. Formats for other reports and logs can be programmed, and the data can be entered via the computer terminals by the operating personnel as needed.

11. *Product blending (mixing) and economics:* Development and application of computer programs for blend (mix) calculation and implementation provide both a higher degree of accuracy and the opportunity to use the most economically advantageous formulation within specification limits.

12. *Personnel (staffing):* Compared with manually operated plants, a reduction of up to 67 percent in the total number of people can be attained for the processing operations (operating, maintenance, clerical, supervisory, and management personnel combined) *if effectively managed.*

13. *Reliability:* With both proper hardware selection and an effective (high-performance) operating team, greater than 99.0 percent reliability can be consistently achieved.

14. *Information access:* The system provides a variety of information access capabilities. First, within the capacity of the particular system, archival storage of information normally accessible to the computer can be provided. Nonprogram operating instructions (start-up and shutdown procedures, for example) can be stored in the computer for recall by and display for the operating people. Information from peripheral operations such as quality control and maintenance can be entered via remote terminals for direct process control file update or instructional access by the operating people. Finally, the process control computer can be networked for information access by other authorized personnel at either the plant site or remote locations.

The potential benefits are both impressive and achievable. Today, most plants could justify the cost of retrofitting with computer process control.

However, the total picture needs to be examined, including the shortcomings and how to accomplish the installation successfully.

Shortcomings.
Scheduled plant start-ups and shutdowns. If not for plant start-ups and shutdowns, computer control operation would imply pressing the terminal start-up key, operating key, or shutdown key to cover all possible needs. Unfortunately, it is not quite that simple. While technically feasible, it is not economically justifiable to size the computer hardware and develop the computer software to cover plant start-ups and shutdowns. Thus, manual procedures are required for most process start-ups and shutdowns, but this does not necessarily imply manual local control of devices in all cases. Devices that are normally controlled by computer programs can be accessed via the computer, but the operator is required to initiate the commands.

Unscheduled shutdowns and subsequent start-ups. While scheduled start-ups and shutdowns are obvious needs, unscheduled shutdowns and subsequent start-ups are potentially more disruptive. Electrical power outages are the best example of unscheduled shutdowns. A skilled operating team can recover from an electrical power outage in minutes, probably faster than with manual control, but if the operating team is not sufficiently skilled in both the control and process system dimensions of the operation, recovery from a momentary outage could take hours.

Initial start-up operation with computer process control. Not to be confused with ongoing plant start-ups (scheduled or unscheduled), the initial plant or process start-up operation is much more difficult with computer process control than with conventional control. Regardless of how thoroughly everything—hardware, software, and process control devices—has been approached, defects will be found and correction required before operation can occur. Nothing operates with computer process control unless every element is working properly. Initially, this reality of computer process control is frightening. Initial start-up is a time of great demand on both the start-up team and the total organization, which requires a high degree of commitment, discipline, patience, and perseverance to prevail. Once the initial problems have been resolved and the operating team has developed the skills to immediately recognize, diagnose, and resolve the ongoing problems that will occur, computer process control is more reliable and beneficial than conventional control.

Skill requirements. While fewer total personnel are required with computer control, the skill requirements are significantly greater than with conventional operation. To achieve the benefits outlined, every member of the operating team must be capable of understanding, operating, troubleshooting, and maintaining all but the most complex aspects of both the computer control system and the process control devices. If not, the installation of computer control will result either in an increase in staffing or in failure to achieve the potential benefits. The last statement raises the argument that if the benefits are as has been outlined, adding a few more highly trained specialists to operate the computer system is probably justified. The fallacy with this argument is that it takes more than a few specialists to successfully operate with computer control. Experience has demonstrated that people assigned exclusively to operate computer systems involving cathode-ray tubes (CRTs) and keyboards incur rapid emotional burnout and become ineffective after a few weeks. Thus, job rotation is an essential requirement, for that reason and others. To perform at peak level the personnel operating in the field need knowledge of the control system, and those operating in the control room need intimate knowledge of the control devices, equipment, and processes in the field. The same is true with respect to maintenance functions. People performing maintenance need to understand both the control and processing systems thoroughly in order to work efficiently and ensure compatibility with ongoing operations. To consistently get the job done with computer control requires a close-knit team of multiskilled technicians, who develop and maintain peak performance by a variety of job assignments and a high level of participation in the management process.

Resources.   As with the application of any new technology, most company staffs do not have an abundance of expertise to independently implement computer-based process control. Consequently, external resources are utilized to some degree, which can range from a complete turnkey project (maximum) to one-person guidance (minimum). The choice should be made by careful analysis of needs and objectives. Some general points should be considered in deciding on external resource use. First, at this stage of development of computer process control, there are no "painless births." In other words, ultimately the end user (the client) must bear some of the pain to master the system and to receive the payback; external resources can carry some but not all of the burden. The end user must *never* abdicate control of the project to the external resource(s), since the probable price paid for abdication of responsibility is protracted and more expensive start-up of the system. Next, as a general rule, maximum use of outside resources will reduce the project time, increase the project cost, and, depending on the thoroughness of preparation by the permanent team, increase the start-up time. Minimum use of outside resources will significantly lengthen the project time, reduce the project cost, and reduce the start-up time.

Dairy Process Control

The previous case study example was helpful in establishing credibility for the various benefits and advantages that can be realized from computer-to be properly addressed in order to achieve successful results. The present case study example will focus more on the "how to" in establishing the technical specifications for a process control system. It contains example process flow diagrams and process instrumentation drawings that are used to specify the control point requirements for a proposed control system applied to a dairy process. The process is designed for the manufacture of fluid milk products of varying fat content, with the excess cream being sold in bulk, as described by Hyde (1986). It is a hypothetical process, not representing any specific dairy plant, which has been provided as a courtesy of Sieberling Associates, Inc. and is based on their collective process control experience within the dairy processing industry.

As explained earlier in this chapter, the first step in specifying a process control system is to prepare process flow diagrams showing all the unit processing operations that need to be controlled and the sequence in which the product moves from one operation to another. A flow diagram of the overall fluid milk process described in this example is shown in figure 3.9. Raw milk is received in tank trucks shown in the upper left corner of the diagram and is pumped to raw storage tanks (RT1 and RT2) shown in the upper right corner. The clean-in-place (CIP) system is shown just below the tank truck receiving area. Raw milk is then centrifuged to separate the cream from the skim, and the cream is pumped to the cream storage tank (CR1) while the skim is pumped to the skim storage tanks (SK1 and SK2). According to the fat content desired in the final product, the appropriate amounts of skim and cream are metered and blended in-stream enroute to the pasteurizer (heat exchanger) and homogenizer prior to storage in finished product tanks (PT1, PT2, and PT3) to await scheduled filling by delivery to the designated fillers (FL1, FL2, and FL3). For purposes of clarity, only process piping (e.g., no CIP piping) is shown. In addition, all major control valves, pumps, metering devices, etc. are shown, identified, and labeled by code number for future specification on related documents.

A master flow diagram such as this provides the process control engineer with the "big picture" that is necessary for gaining a proper perspective and orientation of the system. The next step is to take a more detailed look at each unit operation in order to identify and characterize all the specific control points that will require detailed specification. This is shown in figures 3.10 through 3.13, which serve as examples of detailed process instrumentation drawings for the major unit operations in the process. Control points for the receiving and separating area are detailed in figure 3.10. The standardizing and pasteurizing control points are shown in figure 3.11, while figures 3.12 and 3.13 describe the filling control points and CIP system control requirements, respectively. These detailed instrumentation diagrams are a valuable aid in identifying where various measuring and monitoring functions (feedback control loops, counters, timers, etc.) are needed and become the basis upon which a list can be made of the various control system hardware components that will be needed and their functional specifications.

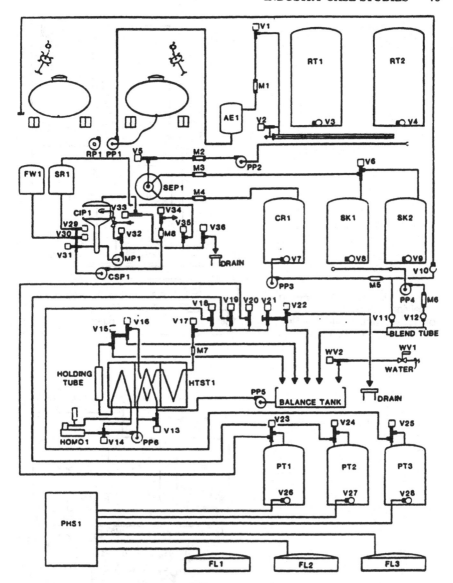

FIG. 3.9.   Process flow diagram of dairy plant producing fluid milk with various fat contents. (Courtesy Seiberling Associates, Inc.)

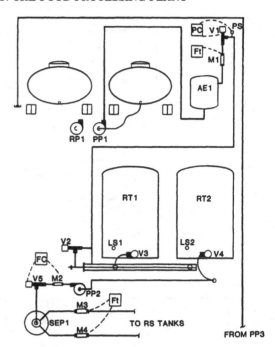

FIG. 3.10. Control points in receiving and separating areas of dairy plant.(Courtesy Seiberling Associates, Inc.)

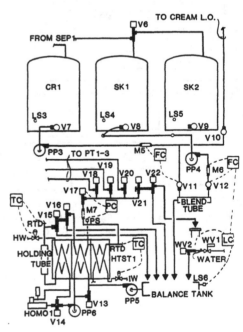

FIG. 3.11. Control points in standardizing and pasteurizing areas of dairy plant.(Courtesy Seiberling Associates, Inc.)

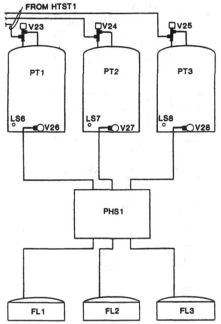

FIG. 3.12.   Control points in finished product storage area and filling line operations of dairy plant. (Courtesy Seiberling Associates, Inc.)

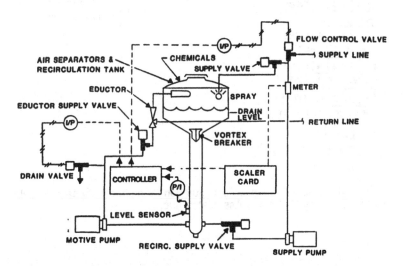

FIG. 3.13.   Control points for clean-in-place (CIP) unit operation in dairy plant. (Courtesy Seiberling Associates, Inc.)

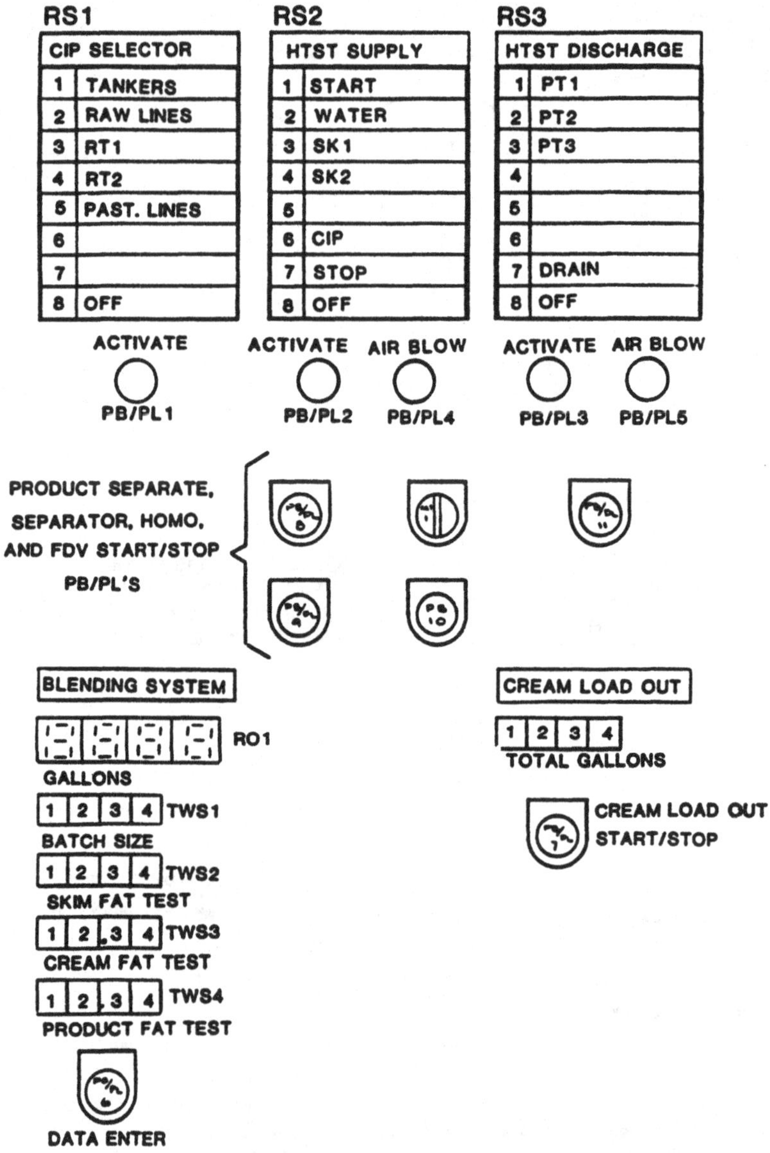

FIG. 3.14.   Operator interface configuration for main control panel in dairy plant. (Courtesy Seiberling Associates, Inc.)

Once all process control points have been specified, it is possible to provide specifications for the type of operator interface that may be desired. These specifications may include the number, location, and type of control panels required; the types might include pushbutton consoles, keyboard terminals with CRT display with or without color graphics, and so on. It is best to provide drawings or sketches of the precise control panel configuration or display screen configuration desired. For the fluid milk process in this case study, three control panels were specified. A main control panel located in a central control room would control the CIP, pasteurization, blending, and cream load-out unit operations, while separate panels located in the receiving area and in the filling room would control all receiving and filling line operations. The configurations specified for these panels are shown in figure 3.14 for the main control panel and in figure 3.15 for the receiving and filling panels.

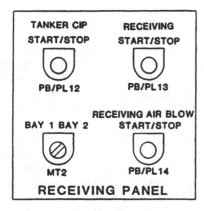

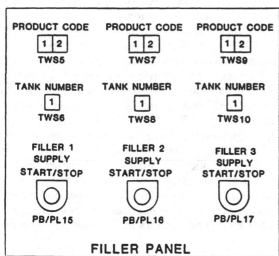

FIG. 3.15. Operator interface configurations for remotely located control panels in receiving area and filling room of dairy plant.(Courtesy Seiberling Associates, Inc.)

The final step in system specification for this case study would be to choose among the various schemes or configurations that were discussed under Control System Configurations. Figure 3.16 shows how control of the fluid milk process would be configured by using a traditional relay-based control system with hard-wire connections for all the process I/O and interpanel communication wiring required throughout the plant. Figure 3.17 shows how the same process would be configured by using a dedicated digital control scheme based on the use of computer-based programmable controllers dedicated to the various unit operations. A centralized control scheme for the same process is shown in figure 3.18, in which a single minicomputer or large-capacity process control module is in control of all operations through communication with remote I/O modules and terminals. Finally, the distributed control configuration for the fluid milk process is shown in figure 3.19 for maximum flexibility and sophistication, as explained earlier in the description of such control systems.

Computer Control in the Distilled Liquor Industry

The next case study example describes the application of computer controls in the liquor industry, reported by Lonberger (1983) for Hiram Walker & Sons, Inc.'s Fort Smith, Arkansas facility. The focus in this application is on production scheduling and automated storage and retrieval, with total integration into the company's management information system.

System Overview.   The manufacturing process at Fort Smith include production of cordials in bulk; bottling of bulk cordials; modification of the packaging of cased goods, for example, gift wrapping and the application of special state bottle stamps or case labels; and order *picking,* that is, the selection of cased goods to meet specific customer orders. There are also associated receiving and shipping functions. All these processes are automated to varying degrees by computerized systems.

The manufacturing, distribution, and facilities monitoring systems employed by Hiram Walker & Sons are highly automated by current standards. These processes are controlled by a hierarchy of computer systems with a number of alternative modes of operation in case any of the systems are inoperative for any reason. In overview, the computer network consists of a Honeywell DPS/8 mainframe MIS (management information systems) computer, 7 Digital Equipment Corp. (DEC) PDP11/34A minicomputers, 10 Allen-Bradley 1774-PLC programmable controllers, 11 Computer Identics Model 60000 laser-beam bar code scanners, and various smaller, more specialized microcomputer-based controllers. In addition to this Fort Smith–based equipment, there are remote computer installations at various contract bottlers, which tie into the Fort Smith DPS/8 for the communication of production and distribution schedule and feedback information. The DPS/8, in turn, is tied into the corporate headquarters in Windsor, Ontario for

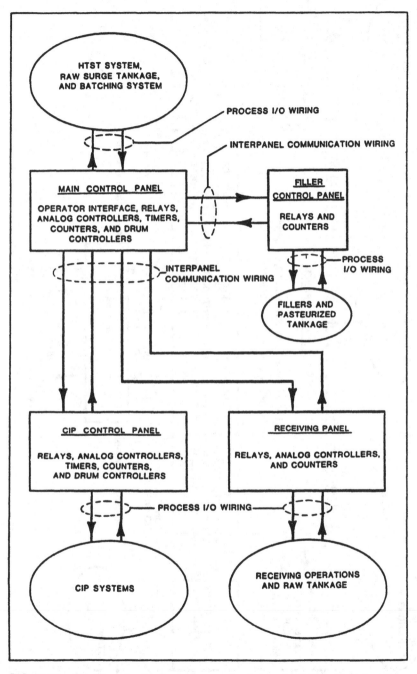

FIG. 3.16.   Traditional relay-based control scheme for fluid milk dairy plant. (Courtesy Seiberling Associates, Inc.)

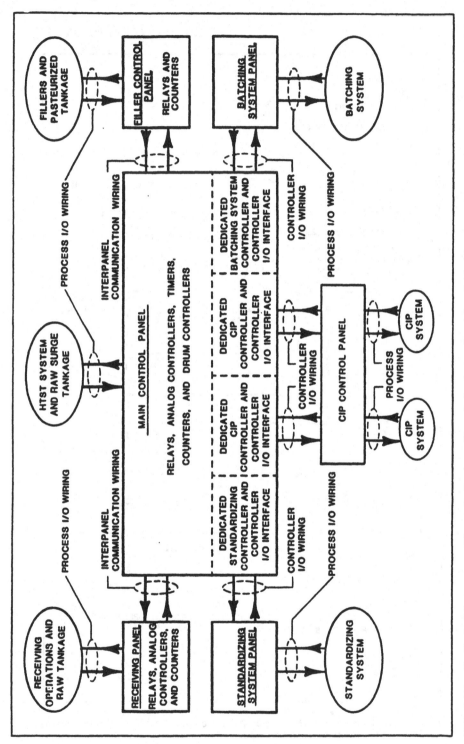

FIG. 3.17. Dedicated (programmable controller-based) control scheme for dairy plant. (Courtesy Seiberling Associates, Inc.)

80

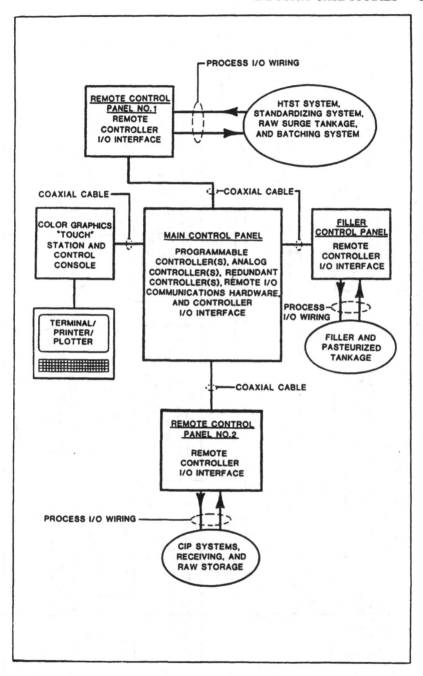

FIG. 3.18.   Centralized computer control scheme for dairy plant. (Courtesy Seiberling Associates, Inc.)

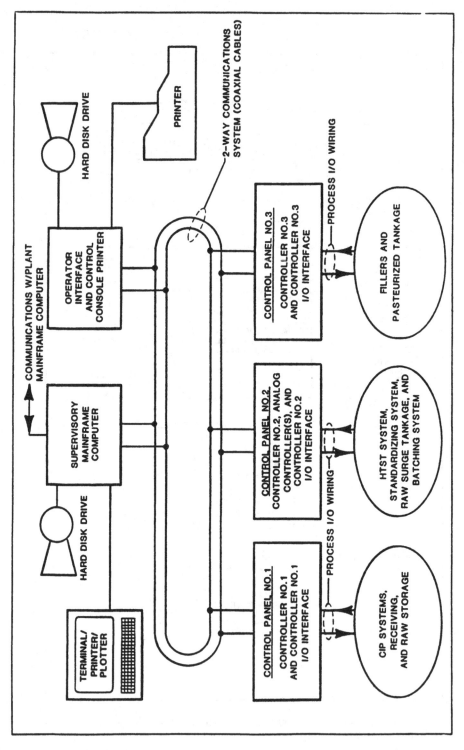

FIG. 3.19. Distributed computer-based process control system for dairy plant. (Courtesy Seiberling Associates, Inc.)

interchange of relevant data with corporate data files. A schematic diagram of this system overview is shown in figure 3.20.

Scheduling of Manufacturing Activities. The bottling and gift packing schedules are driven from sales forecasts provided by the Hiram Walker marketing staff, and the distribution scheduling and scheduling of various product modification processes (state stamping, case coding) are driven from

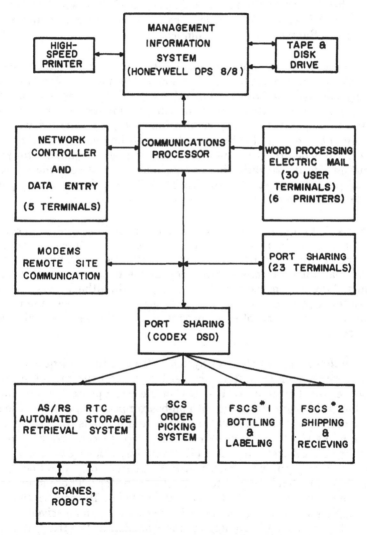

FIG. 3.20. Overview of computer-based process control system configuration at Hiram Walker & Sons, Inc.'s distilled liquor production and distribution facilities. (From Lonberger, 1986.)

customer orders as received into the MIS computer system at Fort Smith. Those schedules involving remote (from Fort Smith) sites are then tele-communicated to the appropriate site, and those schedules to be processed at Fort Smith are similarly communicated upon demand, in batch mode, to the appropriate processing computers, namely, the sortation control system or the factory supervisory computer system. The Fort Smith schedules are developed to an arbitrary horizon, that is, the schedule received may include jobs not due for several days, so that if the MIS computer system is down for a few days, the impact on plant activities will be minimal. If the system is down longer, schedules can be input manually via terminals connected to the factory control computers.

Bulk Liquor Production.   The production of bulk liquors is the least automated process. Some tank utilization and planning information is provided by the MIS computer system; otherwise, the process is accomplished through manual activities, which include the unloading of raw materials (sugar, spirits, flavors), the preparation of highly purified water, the blending of these ingredients, the filtering of the blends, and the settling of the cordials before filtration and transfer to bottling supply tanks. The scheduling of tank utilization is also done manually. Some of these processes are being automated at a Hiram Walker facility in Scotland, and perhaps Fort Smith will be similarly automated in the future.

Process Automation.   The second level in the Fort Smith hierarchy is the complex of seven factory control computers. Only four of these need to be on-line at any given time to support the plant; however, five are typically used to maximize responsiveness. Hence, there is, in general, a backup computer to replace any computer that might fail, although responsiveness may suffer depending on the degree of utilization. These computers are also fed by an uninterruptable power supply (UPS) to prevent disruptions due to power dips.

Automated Storage/Retrieval System.   Two of the on-line and two off-line factory control computers are used to control the plant's automatic storage/retrieval system (AS/RS). One of these computers with a real-time controller (RTC) communicates with on-board microcomputers on the seven stacker cranes in order to store and retrieve unit loads from the high rise storage racks (10 tiers high, 82 columns deep, 14 rows wide). This RTC computer also controls the scheduling and operation of a fleet of 14 automatic remotely guided vehicles (ARGVs), spread over two floors, which move these unit loads to and from the various stops. These stops include 6 shipping and receiving docks, 10 bottling loops, 10 modification areas, 1 dry supplies station, and 3 order picks. The cranes can be driven manually from on board or from a remote operator station at the front of the high rise in the event that the RTC computer is down. The AS/RS manager computer maintains the AS/RS data base (i.e., where loads are and what they are) and gives high-level commands to the RTC computer to move loads (e.g., store a load from truck dock 3 into high rise location 0502110). This manager computer also serves as the interface to users of the AS/RS. If the manager computer is down, the RTC computer can be operated in "semiauto" mode to move loads where required.

The cranes' on-board microcomputer controllers are the lowest-level, and hence the most essential, in the hierarchy of computers controlling the AS/RS. If these computers are not operating, the cranes can still be operated via manual controls, but the demands for positioning of the shuttle relative to the storage bays make this alternative impractical for supporting processing at anywhere near the required design rate.

Hiram Walker's ARGVs (called "Robocarriers" by Eaton-Kenway) have no on-board computers, although later models do. Commands to start, turn, center, uncenter, or pick up or deposit loads are communicated directly from the RTC via a serial data link interfaced to the ARGVs through antennae buried in the floor. The serial data is carried via a standard 20-ma current loop link from the RTC computer to BSK (block controller) interfaces, which service various parts of the pathway. These BSK units convert the 20-ma signals to approximately 200-ma levels (modulated and unmodulated), which are then fed to the floor antennae. The ARGVs have front- and rear-mounted antennae for transceiving information. The data transmissions take place in three frequency bands, which are approximately 4, 7.5, and 9 KHz. One frequency band is called the *tracking frequency*. The ARGV will not proceed unless this frequency is present, and it will proceed only in the direction at which this frequency is strongest. Another of these frequency bands communicates data from the ARGV to the RTC, for example, data such as "table high," "load on table," "centering cones up (or down)." The third frequency band is used to communicate commands to the ARGV from the RTC computer, such as "raise (or lower) centering cones," "raise (or lower) table," and "go." Floor-mounted metal plates tell the ARGVs where to speed up (slow, medium, fast) and slow down, as well as where to stop for further instructions from the RTC.

Order Pick–Sortation Control System.   The computer hierarchy for the order picking application goes from the MIS computer (for scheduling) to the sortation control system (SCS) computer (for processing) to two programmable controllers (for actual machine control under supervision of the SCS computer).

This MIS computer creates the schedule of order pick jobs (called *runs*). These jobs consist of a demand to create a collection of shipping cubes (i.e., unit loads of palletized cases), which contain the cased goods for specific customer orders. The job schedule contains the accumulated ordering requirement for all shipping cubes in the job. The ordering sequence from the AS/RS is scheduled so as to enhance system efficiency by ordering products for the cubes using the fewest number of different products before those that use a greater variety. This procedure allows for some cubes to be in the tier sorting and palletizing stages while the cases for the others are still being accumulated; hence, at the end of an 18-cube job, perhaps only 3 or 4 cubes might be waiting. Also, this early completion of cubes frees up the pallet accumulation lanes for the induction of cased goods for another pallet.

The SCS computer provides the interface to the loop supervisor, who can thereby release jobs in the desired order and can monitor the progress of jobs via a real-time display of order picking activities. The SCS computer provides this same information to the operator at the sortation lanes, alerts the operator to problems as they arise, and has provisions for correcting problems (e.g., a mismatch between the computer's idea of what cases are on a given lane versus what is actually on that lane, such as might arise in the event of a case jam).

In addition to this operator interface function, the SCS computer reads the bar code label data as transmitted by a bar code scanner and uses this data to decide what type of case is present at the scanner and into which sort lane this case should be diverted. The SCS computer then communicates this sort lane data to the programmable controller that operates the sortation equipment; this programmable controller sends the case to its appropriate lane and then reports back its success (or failure) to the SCS computer. This data communication is carried on via a serial-data, 20-ma current loop channel from a DEC DZ11 port to an Allen-Bradley 1774-CI2 computer interface card with a 20-ma RS232 converter located near the 1774-CI2. This sortation process can be accomplished by using manual keyboard/display units if the SCS computer is down, but the processing rate is substantially reduced in this mode (to about 50 percent or less of maximum). The system cannot run without the programmable controller. As noted, there is backup computer capability to minimize any requirement for manual operations.

The SCS computer also automatically communicates storage and retrieval requests to the AS/RS computer system so that the necessary input goods are retrieved for the shipping cubes (pallets) to be created and all unused goods, as well as the finished shipping cubes, can be stored back into the AS/RS high rise without manual intervention. This SCS-to-AS/RS computer link involves two directly coupled DZ-11 serial data ports on each computer. One link is used to communicate storage and retrieval information, the other for the AS/RS to inform the SCS of the delivery of incoming goods. This storage and retrieval information can also be communicated directly via AS/RS terminals in case the SCS computer link is inoperative.

Tank Farm Storage of Frozen Concentrated Orange Juice

The following case study will serve as a second example of computer control in an automated storage/retrieval system but will deal with the bulk storage of liquid product in huge silo tanks on a tank farm storage facility rather than with storage of packaged goods in a warehouse. The highly seasonal nature of orange juice processing requires that large amounts of concentrate be stored and used as needed throughout the year. The tank farm described in this case study is part of the citrus processing facility of Orangeco of Florida, Inc. at Bartow, Florida. The control system application is based on a case history presentation by Dorai (1986) of the Foxboro Company.

Process Description.   As shown in figure 3.21, oranges are weighed, graded, and placed in storage bins on arrival at the processing facility. Later they are sent to the juice extractor for crushing, and the extracted juice is pumped to holding tanks and then on to concentrators (evaporators) for water removal. The concentrate is chilled and pumped to the tank farm for chilled storage. On demand, the concentrate is pumped to blenders for the addition of water and/or fresh juice, natural oils, and aroma. The juice is now pumped to the packaging facilities for filling and freezing and finally to a refrigerated warehouse.

Before the installation of the tank farm and the associated computer control and management system, Orangeco used the conventional method of storing orange juice concentrate in plastic-lined, 55-gal drums. The drums were filled with concentrate, sealed, and then palletized, four to a pallet, and stored two pallets high in a freezer. The maximum possible storage

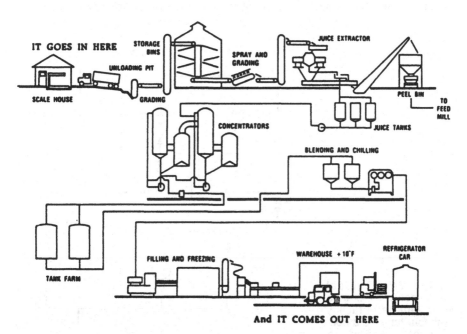

FIG. 3.21.   Process flow diagram for production of frozen concentrated orange juice. (Courtesy The Foxboro Company.)

capacity in the freezer with this method of storage was limited to 1.1 million gal.

The freezer was converted to a tank farm storage facility in the summer of 1980. The tank farm consists of 10 pods (groupings of tanks). Pods 1 through 4 and 6 through 9 each have four tanks with a per tank capacity of 105,000 gal. Pods 5 and 10 have three tanks each, with a combined capacity of 420,000 gal from the six tanks. This translates to a total storage capacity of 3,780,000 gal spread over 38 storage tanks in the farm.

The details of a typical pod are shown in figure 3.22. Each tank is provided with a hydraulically actuated 6-in inlet butterfly valve, a similar outlet valve, and a rotary pump. This pump is driven by a variable-speed hydraulic motor and is used to withdraw the concentrate from any of the tanks in the pod and transfer it to an outbound load cell tank. All concentrate brought into or taken out of the tank farm is weighed in tanks mounted on electronic weigh cells. The details of the inbound-outbound load cell tanks are shown in figure 3.23. Three hydraulic pumping systems supply the hydraulic fluid under the pressure required for the operation of valves and hydraulic motors throughout the tank farm. These hydraulic "power packs" are energized when required for operation.

It is important to note that there are numerous quality variations to contend with, in addition to the quantities, in the various tanks. The concentrate's quality could vary from tank to tank in terms of color, solids, Brix reading, acid, oil, pulp, flavor, defects, and temperature. Management must know these quality details of the concentrate in the inventory.

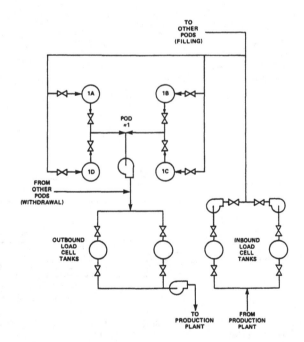

FIG. 3.22. Typical "pod" setup for orange juice tank farm facility with four tanks to a pod. (Courtesy The Foxboro Company.

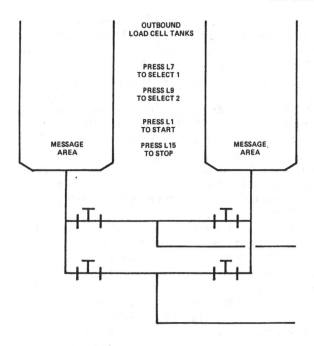

FIG. 3.23. Typical load cell tanks used for controlling product withdrawal from tank pods in orange juice tank farm storage system. (Courtesy The Foxboro Company.)

Control Level Options. In choosing the best control scheme for the tank farms facility, the four levels of control system sophistication discussed below were considered.

Manual Control. In its simplest form, the management strategy could have been to install a level gauge on each of the 38 tanks in the farm. Since the accuracy of level gauges is very low, load cells on each tank may have to be considered. One pump should be installed to receive concentrate into the tank farm, and the piping manually connected to the desired tank. Similarly, on withdrawal, the withdrawal pump should be moved to the desired tank and, with the manually connected piping, the flow directed where desired. The quantities involved should be recorded manually, and the accounting department should prepare the necessary reports and other documentation. It is questionable as to whether this strategy would have helped management meet its objective of eliminating the tank farm as a bottleneck and obtaining accurate and usable knowledge of inventory data at a reasonable cost. Probably there are still some citrus processors who follow this strategy which in essence is a strategy of larger drum storage.

Traditional Relay Control. Another strategy could have been to install remotely actuated inlet and outlet valves on each tank, grouping fewer tanks into a pod and providing a withdrawal pump for each pod, and to have a load cell arrangement on separate dual receiving tanks so that one could be filling while the other empties into one of the tanks in the farm. A similar arrangement could be installed for withdrawal. The data involved in such an operation are recorded manually while keeping a running inventory. All data involved are manually processed. The control function of the pumps and valves could be automated by using conventional relay logic, and the control panel could have a switch for each inlet and outlet valve, a total of 76 for the 38 tanks. Additional switches for each withdrawal pump on the pods would have numbered 10. Level indication of the 38 tanks could be centralized in the same control room, and additional alarm indications of level could be incorporated. The operators could manually accomplish the alternate filling-emptying operations on receiving and withdrawal, or this function could be automated by using conventional relay logic. The selection and sequencing of tanks could be accomplished by the supervisor by studying all records.

This strategy would result in a large and complex control room, with a high resultant cost. The decision process would have required supervisory assistance. With the manual data processing and the complex operator interface, it is doubtful that the original objectives could have been met, even with the resultant high capital and operating cost. Actually, this strategy has been adopted by some citrus processors and, once again, has not realized the full benefits of the tank farm installation.

Dedicated Control with Programmable Logic Controllers. PLCs offer improved flexibility over hard-wired relay logic systems and have proved to be reliable. The strategy could have taken advantage of these developments in PLCs by adapting them to the design of the tank farm control and management system. This would offer the advantages stated earlier over the conventional relay logic system. The main disadvantage would be the operator-machine interface constraints—since PLCs are designed for the dedicated function of relay logic replacement, they are not suited to handle the information requirements of the operator and the supervisor. Obviously, this strategy does not measure up to the requirements established earlier.

Centralized Computer Control. Further analysis showed a need to apply the solid-state technology to both control and manage the tank farm. It was logical to combine the control capability and information processing capability into one solid-state system that could control and manage the tank farm. Study of the available technology presented certain options to meet this strategy, one of which was to use PLCs to control the tank farm and interface with a supervisory computer for data gathering and informational processing. Study of the available technology also showed the availability of many low-priced computer hardware devices. One strategy was to adapt such hardware to control and manage the tank farm.

Another option would be to use a computer specifically designed for process control and information processing, that is, a process control computer. The preferred strategy was to combine process control and information processing into one system and to employ the computer in a centralized control system configuration, as shown in figure 3.24.

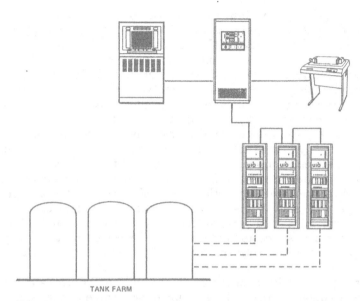

FIG. 3.24.   Centralized computer control system configuration for orange juice tank farm storage facility.(Courtesy The Foxboro Company.)

System Operation.   Since a centralized control system configuration was selected for this application, system operation was centered around the operator interface with the computer. This consisted of a color graphics display terminal with keyboard, which would allow the operator to call up any one of a number of displays controlling various system components. As an example, the operation of withdrawing from the tank farm will be considered. Only limited blending can be achieved because the load cells are used to measure the product withdrawal from each tank; thus, withdrawal may be made from only one tank at a time. In a withdrawal operation, the product will be withdrawn by rotation in the sequence that the tanks are entered on the WITHDRAWAL display. The maximum number of rotations is five. This will provide a reasonable blend in the load cell tank.

The operator must enter the number of tanks to be withdrawn (minimum one, maximum eight tanks) on the withdrawal display. Next, the tank names and the net weight to be withdrawn from each tank must be entered. After each tank name is entered, the computer will check if that tank is available. If the selected tank is busy (involved in a filling operation), the computer will display a message on the withdrawal display and allow another tank selection. After the weight to be withdrawn from a tank is entered, the computer will check its inventory to ensure that a sufficient amount of product is in the tank; if not, a message will be shown on the withdrawal display, and the operator will be allowed to enter a smaller weight.

The computer will perform all cursor control. As each entry is completed, the cursor will be moved to the next location. In this example the next item that must be entered is the number of rotations the computer is to use to fill the load cell tank, and any integer from 1 to 5 will be accepted.

The last entry is the load cell tank selection. If the load cell tank chosen is busy (pumping to production), a message will be shown on the display, and the operator will be allowed to make another selection. If the selected load cell tank has some product in it but is not full, a message will be displayed. An operator wishing to have the current batch added to the product already in the tank will proceed with the load cell tank selection, and the data entry procedure will continue. If the selected load cell tank is available, the computer will sum the requested weights to determine if the load cell tank will accept the requested batch size. If the batch size is too large, a message will be shown on the display, and the data entry procedure must be started again at the top of the withdrawal display. When the computer has accepted the load cell tank selection, it will display an appropriate message with instructions for starting the withdrawal procedure.

The computer will calculate the total weight withdrawn and pounds of solid withdrawn by using the calculated Brix value in the tank and the change in load cell weight while that tank is active. The Brix value for the tank is calculated from the pounds of solid and net pounds in that tank (data from inventory). These calculations are performed and the tank inventory is updated each time a tank has completed its segment of a rotation.

When the withdrawal operation is started, a message will be logged. After each segment of a rotation is completed, a message will be printed indicating how many pounds have been withdrawn from that tank for the current withdrawal operation. When the withdrawal operation is completed or aborted, a message is printed indicating that the operation has stopped and why. If, at any time during the withdrawal operation, one of the load cell inputs is bad or if there is a deviation of more than 10 percent between the highest and lowest load cell weights, the withdrawal operation will be aborted with a printed message on the system terminal.

The original purpose of the computer system was not only to enhance the operation of the tank farm but also to provide timely information to management. In addition to all the previously mentioned operational information continuously available, three separate reporting requirements were established. Two of the reports are hard copy only and are printed on the system terminal exclusively on a periodic basis. The third report is printed either on the operator's console or on the system terminal and is available any time on request.

Batch Sterilization of Canned Food

This final case study example will describe the development of a commercial system for control of batch sterilizers (retorts) in a food canning plant using a centralized microcomputer-based control system. The system as described by Brown (1985) consists of three duplicate centralized control systems to automate the complete production cycle of 24 batch retorts in a food canning plant operated by Hillsdown, Ltd. in the United Kingdom. Details of system hardware and software are given that were not provided in previous case study examples.

Performance Requirements. The objective of the control system, in addition to automation, was to improve control accuracy at strategic points so as to ensure a higher degree of repeatability between batch processes, to improve the accuracy and level of process documentation and to improve general process efficiency in terms of energy management, and operator efficiency. The basic sequence of events that form the retort cycle is reasonably standard throughout the industry, although there are minor differences due to differences in the availability of services in some retort installations and there are occasional special requirements for coping with unusual packaging or products. The cycle used is typical of the "standard" retort cycle. First, cans are loaded by machine into large crates, containing up to 1500 cans (depending on can size), and the retort is manually loaded with three crates and then has its lid bolted down. The heat treatment process is divided into the seven stages outlined below.

1. *Preheat:* The retort and cans are warmed up to 99°C by using saturated steam, while allowing condensate formed on the cans to drain away to avoid submerging the lowest cans later in the process.
2. *Venting:* Air is removed from the retort to ensure a pure steam environment so that there is no temperature distribution within the retort and thus to ensure that each can is subjected to precisely the same heat process.
3. *Bringing up:* The temperature of the retort is raised from that achieved at the end of the vent period to that required for sterilization.
4. *Processing:* The retort environment is maintained at exactly the sterilizing temperature for a prescribed duration to ensure that a sufficiently large time-temperature integral is achieved to provide the required degree of sterility.
5. *Pressure-cool:* The pressure and temperature within the retort are reduced at a rate that will not cause damage to the cans (they will have developed a higher internal pressure during stage 4). It is usual to create an overpressure, higher than the pressure required to support saturated steam at the current retort temperature, in order to ensure that the pressure differential across the can wall does not cause either the can lid seal to be broken or the can walls to be damaged.
6. *Atmospheric cooling:* The temperature of the product is reduced rapidly to below 60°C to avoid the risk of incubating any surviving bacteria (and thus increasing their numbers) and to complete the product's heat treatment.
7. *Emptying:* The final step is to remove any remaining cooling water from the retort and to ensure that the temperature and pressure

within the retort are at safe levels to allow the retort lid to be opened and the processed cans to be removed.

Hardware Requirements. The arrangement of pipework, valves, and sensors is shown in figure 3.25. All valves used, except the steam valves, are simple on/off valves, driven pneumatically by air supplied via small electrical solenoid valves, which are actuated by the control system. The steam valves are fully modulating control valves, again driven pneumatically but actuated via current-to-pressure (I/P) converters running on four 20-ma current loops, generated by the control system. The steam valves have "equal percentage" plug characteristics and have been sized to supply both the large flow rates required during the venting stage and the smaller more carefully controlled flow rates required during the processing stage. The speed of response of the steam valves and the accuracy with which they can be positioned are important in this batch situation, since it is important to achieve the process temperature quickly and accurately to ensure the correct heat treatment for the product.

Temperature within the commercial retorts was sensed by using platinum resistance thermometers (PRTs) fitted with current-loop transmitters.

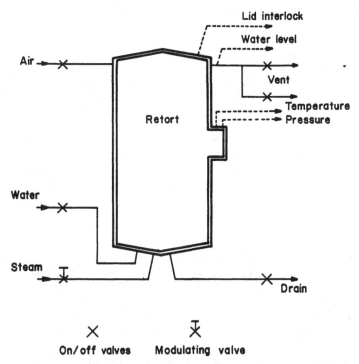

FIG. 3.25. Arrangement of piping, valves, and sensors required on batch retort for heat sterilization of canned foods. (From Brown, 1985.)

The thermometers were installed in a continuously vented instrument pocket located on the side of the retort wall. The transmitters were fitted into the connecting heads of the thermometers and permitted simple two-wire connections to the remainder of the measuring system. The configuration, shown in figure 3.26, allowed the temperature sensed by the PRT to be displayed locally, recorded remotely, and used as the input signal to the microcomputer.

The PRTs were selected to discern temperature with an accuracy of better than ±0.25°C. The current loop transmitters were accurate to better than ±0.1°C ±0.05 percent of reading, and so the total error in temperature sensing was less than ±0.4°C at 120°C. The accuracy of the microcomputer in discerning the value of the incoming temperature signal was better than ±0.1°C at 120°C, and so the complete system accuracy was better than ±0.5°C. The local temperature display and remote temperature recorders each have their own accuracy limitations; total errors were better than ±0.6°C and ±0.55°C, respectively.

Pressure sensing employed a similar current loop arrangement. The sensors chosen were strain-gauge diaphragm types with integral current loop transmitters. The instruments were located alongside the PRTs. The individual retort pressure was displayed locally and transmitted to the

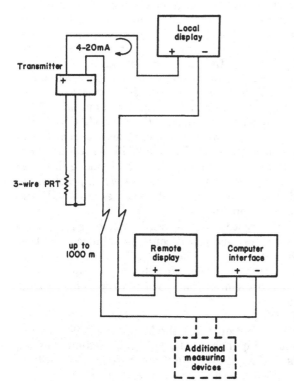

FIG. 3.26. Current loop arrangement for transmitting 4- to 20-MA signal from temperature sensors through computer-based retort control system. (From Brown, 1985.)

microcomputer, no additional facilities were provided to record pressure. The accuracy of pressure measurement was better than ±0.7 psi at the microcomputer and at the local display.

A water level sensor was positioned so that it gave a digital signal when the topmost layer of cans in the retort was covered by at least 3 in of water. An electrical interlock was fitted to each retort lid as a safety feature. Opening the lid caused the drain valves to open and caused all other valves to close. In addition, the LID OPEN signal was transmitted to the microcomputer and used as a software reset. The whole commercial system was required to have a manual operation facility to be used in the event of a microcomputer failure, sensor failure, or local minor disaster and to provide a manual means of recovering from a situation in which the retort process parameters were out of limits because of a local service supply problem or a mechanical fault in the retort. In addition, a graphic display panel and operator interface were required for each retort. A separate and special interface was designed and constructed to provide a junction between the retort hardware and either microcomputer or operator commands, to provide the logic to operate the manual operator interface and mimic panel, and to provide the appropriate logic required to perform safety functions locally at the retort and thus reduce the dependence on microcomputer response in an emergency.

The geography of the factory and the style of production allowed the production management staff to arrange for the flow of product to be directed from specific preparation lines to one of the three banks of eight retorts. Consequently, it was decided to use three duplicate MACSYM 350–based systems from Analog Devices, Ltd., one for each bank.

Each MACSYM 350–based system was arranged in a centralized control style, and each item of plant hardware, including the mimic panel, was connected individually to the analog and digital input and output (ADIO) interfaces installed in the MACSYM 200 chassis. Both the MACSYM 150 chassis and the MACSYM 200 chassis were located in the same room away from the production area and environment. Figure 3.27 illustrates the basic system architecture. Note that a single printer was shared among the three systems by multiplexing its input manually. This caused no problems in time sharing because production records were only required once daily and were obtained outside of production time.

The style of the system's architecture was carefully considered to ensure that the independent operation of each retort could be achieved and that the system *service rate* (the rate at which the MACSYM 350s could meet the requirements of the retorts) was sufficient to achieve the required temperature control accuracy and to ensure rapid reaction to operator instructions and emergency situations.

The choice of plant hardware and the use of remote mimic/operator panels dictated the selection of the ADIO required in the MACSYM 350s. In total, each retort required two analog input channels, one analog output channel, five digital input channels, and eight digital output channels. In addition, three analog input channels were required for each bank of retorts to report

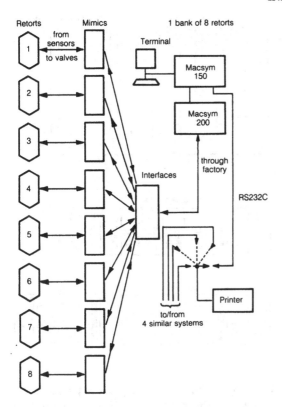

FIG. 3.27. System architecture for centralized computer control of eight batch retorts in commercial food canning plant cookroom operation. (From Brown, 1985.)

the pressure in the air, steam, and water service manifolds. Each MACSYM 200 therefore contained 11 interface boards connected to 131 ADIO channels.

The incorporation of an electronic logic board into each mimic/operator panel greatly reduced the number of digital input and output channels that otherwise would have been necessary to communicate with each individual function of the operator's panel. It also simplified the selection between AUTO and MANUAL plant operation modes, provided a convenient place to condition the signals emanating from pushbuttons (to remove switch bounce, etc.) and provided local independent facilities for safety interlocks and emergency procedures.

Software Requirements. Figure 3.28 illustrates the fundamental software architecture. The seven stages of the process cycle were supported by a set of subroutines within the control software. Each of these subroutines contained appropriate statements to implement its control functions. Each retort was assigned a specific data array, containing: a representation of the state of all its valves, pushbuttons, and interlocks; current values

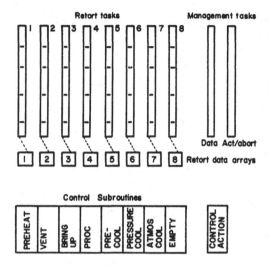

FIG. 3.28. Software architecture for centralized computer control system of multiple batch retort operations. (From Brown, 1985.)

of temperature and pressure; current values of other process parameters; values of set points required to define the process (including two pairs of tolerance limits for each set point); constants to be used in the control algorithm associated with its steam valve; and a number of flags and other items associated with the retort's progress through its cycle. Each subroutine in the *control* set acted on the data in the *retort array*. After each action, a further subroutine acted upon the retort array data and implemented the required ADIO functions and hence operated on the retort. In addition, part of a set of *support* subroutines was used to update the retort arrays according to changes in the real world by: performing data acquisition and conditioning; providing cover for emergency situations; checking current values for process parameters against set points and then checking their compliance with the appropriate tolerances; and initiating corrective or abortive action, whichever was required.

Other subroutines in the support set provided the necessary internal housekeeping duties, controlled a local graphics display of the current state of all eight retorts, provided data updates, controlled data collection, conditioning, compaction, and storage of the MACSYM disks on a regular basis (at 1-min intervals) and monitored the ACTIVE/ABORT commands at the operator panel in order to initiate a retort cycle on demand.

The multitasking ability of the MACBASIC software package allowed each retort to have its own task and so enabled each retort to be operated individually. In addition to the eight retort tasks, two other tasks consisted of high-level data management and supervision of retort active/abort operations. Each task had a number of the support set subroutines available.

The principles associated with the use of a common set of control subroutines allowed each retort to be treated in a similar way; however, the use of retort-specific data arrays ensured that each retort received the appropriate action it required. The use of independent high-level retort tasks ensured that each retort could be operated independently from the others and allowed the retorts to perform different processes at different times while maintaining a high degree of process plant integrity. The use of a separate support set of subroutines and a pair of high-level service tasks ensured that the retort tasks ran smoothly, provided an up-to-date graphic display, and maintained accurate and reliable data logging onto disk. In all, some 74k bytes (a measure of the amount of memory required) of software was written, containing over 1300 lines of commands, a large portion of which allowed production staff an opportunity to work on the process parameters, both in preparation for a process session and while the system was performing its control functions.

The principle of centralized control required centralized software. This example was complicated because of the speed required to achieve process accuracy and reproductibility, but the memory capacity was sufficient to contain all the control software. Other situations may place excessive software loads on the central microcomputer, which may become larger than the system's capacity for ADIO. In these circumstances some degree of control distribution may be necessary simply to relieve software pressures.

Canned food sterilization in batch retorts will continue to appear as a common case study example in Chapters 4, 5, and 6, dealing with on-line control of unit operations, process modeling and simulation, and process optimization, respectively. Although other unit process operations will also be included among the case study examples in these chapters, canned food sterilization will be the only process unit operation that is common to all chapters. By following this one case study example in progression from Chapter 3 through Chapter 6, a full appreciation can be obtained of the advances that are possible with computer technology and engineering mathematics in the design and control of food processing operations.

REFERENCES

BROWN, G.   1985.   Microcomputer controlled batch sterilization in the food industry. Chemistry & Industry 3 (June).

DORAI, G. D. J.   1986.   Computer assisted management of tank farms—a case history. Presented at Short Course on Computer Control of Food Processes, University of Wisconsin Extension, Madison.

HYDE, J. M.   1986.   Dairy Process Control. Presented at Short Course on Computer Control of Food Processes, University of Wisconsin Extension, Madison.

HYDE, J. M., and CLEM, L. W.1985.   Effective process control scheme selection. American Society of Agricultural Engineers, Food Engineering News: Oct.

LONBERGER, J. W.1986.   Computer control in the liquor industry. Presented at Short Course on Computer Control of Food Processes, University of Wisconsin Extension, Madison.

SHAW, W. T., and McMENAMIN, J. F.1984.   Computer Control of Batch Processes. EMC Controls, Inc., Cockeysville, MD.

SMALLWOOD, N. J.1985.   Productivity management in vegetable oil refineries. Presented at the 76th Annual Meeting of the American Oil Chemists Society, 5– 9 May, Philadelphia.

4

On-Line Control of Unit Operations

This chapter describes computer applications in the control of specific unit operations that make up any one stage of processing.

GENERAL CONCEPTS

A *unit operation* can be defined as the operation of a piece of equipment or equipment system that accomplishes some specified function on the product being processed and is typically labeled by the function it accomplishes. Some of the unit operations commonly found in food processing are heating and cooling, mixing and blending, pasteurization, sterilization by thermal processing, freezing, evaporation, dehydration, fermentation, distillation, extraction, and separation. For any one unit operation, a wide variety of different types of equipment and systems can be used, depending on the specific nature of the product being processed.

Traditional control of such unit operations consists of maintaining specified operating conditions that have been predetermined from product and process development research, for example by controlling the liquid level in a tank, the product temperature at the outlet of a heat exchanger, or the time and temperature of a batch cook. As explained in Chapter 3, computer-based control systems can easily replace traditional automatic controllers in executing the control loop logic for these operations when properly interfaced with sensors, valves, and switches on the operating equipment or system. When these unit operation control systems are integrated with the centralized or distributed computer control system throughout the plant, the entire plant process can be viewed to be under computer control.

At this level of computerization, however, process control is still limited to maintaining prespecified processing conditions at each unit operation. Sometimes unexpected changes that occur during the course of a unit operation or at some point upstream in a processing sequence can make the prespecified processing conditions no longer valid or appropriate, leading to off-specification product, which must be either reprocessed or discarded at appreciable economic loss. Such situations can be of critical importance in food processing operations because the physical process variables that can be measured and controlled are often only indicators of complex biochemical reactions that are required to take place under the specified process conditions. A classic example is the control of time and temperature to achieve some specified level of bacterial inactivation in the pasteurization or sterilization of a product. Other examples include the control of time, temperature, and flow rates in achieving specified levels of water removal in evaporation or dehydration processes or of biochemical reactions in a fermentation process.

Food engineers with advanced training in heat and mass transfer and reaction kinetics are capable of developing mathematical models that can be programmed in computer software to simulate these processes. When unexpected changes occur in the course of a process, these models can predict the outcome of the reaction as a result of these altered conditions and can be used to calculate and implement downstream changes in process conditions to compensate for upstream deviations, so that the desired final process result is achieved. When these computations and corrective actions can be made rapidly without interrupting the process as it proceeds in real time, the process is operating under intelligent on-line computer control. By using mathematical models for process simulation in this way, the computer can keep track of exactly what is going on during the process as a result of the actual process conditions experienced and can continually adjust future conditions to ensure that the end processing result is achieved.

Computers have far greater computational capacity than is required for executing simple control loop logic. Therefore, the use of such simulation models for on-line computer control of unit operations is a means for taking maximum advantage of computer-based control systems. The case studies presented in this chapter are intended to illustrate the benefits of such online control with process simulation when applied to unit operations involving thermal processing, aseptic processing, multistage evaporation, fermentation, and dehydration. Although the use of mathematical models for on-line control applications will be illustrated for these unit operations, no attempt will be made here to describe their development. The development of mathematical models for process simulation and optimization will be described in Chapters 5 and 6.

THERMAL PROCESSING OF CANNED FOODS

Because of the important emphasis that must be placed on the public safety of canned foods, processors must operate in strict compliance with the U.S. Food and Drug Administration (FDA)'s low-acid canned food regulations. Among other things these regulations require strict documentation and record keeping of all critical control points in the processing of each retort load or batch of canned product. Particular emphasis is placed on product batches that experience an unscheduled process deviation, as when a drop in retort temperature occurs during the course of the process. In such a case the product will not have undergone the established scheduled process and must be either fully reprocessed or set aside for evaluation by a competent processing authority. If the product is judged to be safe, batch records must contain documentation showing how that judgment was reached. If the product is judged unsafe, it must be fully reprocessed or destroyed. Such practices are costly, and processors tend to operate with higher retort temperatures and longer process times than those called for in their established process specifications in order to minimize the frequency of process deviations. At the same time, however, processors recognize that higher retort temperatures and longer process times tend to have an adverse effect on product quality and would like to minimize such overprocessing as much as possible.

Recognizing these difficulties, suppliers of equipment and services to the food industry have begun to offer microprocessor-based electronic systems for the computer control of retorts (Getchell, 1980). These systems promise to be helpful in minimizing the occurrence of process deviations in retort operations and in reducing the cost of record keeping. Ideally, however, these systems should also be capable of rapid evaluation, on-line corrections, and printed documentation of any process deviation that may occur while the process is still underway. This would allow for the release of all product batches on schedule, with full documentation in compliance with regulations.

The purpose of the following case study is to show that these goals are achievable by making use of a numerical computer model that was developed to simulate the thermal processing of conduction-heated canned foods. This model was first developed by Teixeira et al. (1969) for application to cylindrical cans, and later modified by Manson et al. (1970, 1974) for adaptation to rectangular and pear-shaped containers, as well as to discrete particles suspended in liquid foods (Manson and Cullen, 1974). Because these models required the core memory that was then only available in large mainframe computers, they were viewed as research tools for use off-line in evaluating and optimizing thermal processes. Many of the small, inexpensive microprocessors of today have much of the capability of older mainframe machines. This means that a single microprocessor can be programmed with one of these numerical models as an integral part of the control logic used in the real-time on-line control of retorts.

On-Line Control Logic

In controlling thermal processes, the objective is to meet the designed level of bacterial sterilization F_o^d for the process, irrespective of any retort temperature variation $T_R(t)$, and with a minimum of overprocessing. The lethal effects of thermal processing are achieved during the heating as well as the cooling times (t_h and t_c, respectively). Thus, the objective function in thermal processing is to minimize the function

$$F_o(t) = \int_0^{t_h} 10^{\frac{T-250}{z}} \, dt + \int_{t_h}^{t_c} 10^{\frac{T-250}{z}} \, dt \qquad (4.1)$$

subject to the constraint

$$F_o(t) \geqslant F_o^d \qquad (4.2)$$

Here $F_o(t)$ is the accumulated amount of sterilization (lethality) at any time t, the first integral is the contribution to the sterilization $F_o(t)$ from heating, and the second integral is the contribution from cooling. The temperature $T(t)$ is taken to be the temperature of the slowest heating point in the product. This is so that when the design F_o (i.e., F_o^d) is satisfied at this point, all other points in the product have also been satisfied. For conduction heating food in a cylindrical can $T(t)$ is the temperature at the geometric center of the can at time t.

This transient temperature $T(t)$ is a function of the transient retort temperature $T_R(t)$, the dimension of the can R and H, and thermal diffusivity α of the product. Thus, symbolically

$$T(t) = f[T_R(t), R, H, \alpha] \qquad (4.3)$$

For conduction heating in a cylindrical can, the function f is the heat conduction equation in cylindrical coordinates

$$\frac{1}{\alpha} \frac{\partial T}{\partial t} = \frac{\partial^2 T}{\partial z^2} + \frac{1}{r} \frac{\partial T}{\partial r} + \frac{\partial^2 T}{\partial r^2} \qquad (4.4)$$

with the boundary condition

$$T_{\text{boundary}} = T_R(t) \qquad (4.5)$$

Equations 4.1 through 4.5 describe the thermal processing of a product in a cylindrical can heated by conduction.

For a given can size and product, the two variables that can possibly be controlled are temperature $T_R(t)$ of the heating medium (saturated steam or hot water under pressure in the retort) and the heating time t_h. Cooling is generally done with available water at ambient temperature, which makes the contribution to F_o from cooling (given by the second integral in equation 4.1) immune from further control once the cooling process is under way. Also, both the heating and cooling temperatures can only be applied to the boundary of the product.

In the case of a process deviation, the temperature of the heating medium may go through unexpected variations beyond the capacity of the temperature controller. Thus, in reality there is only one variable that can be controlled, namely the heating time period t_h. The problem in thermal processing is to specify t_h for arbitrary variations in $T_R(t)$.

One possibility is to measure the temperature at the can center $T(t)$ with a thermocouple in a test can while the process is underway on-line. This would eliminate the need for function f in equation 4.3. In this approach the measured $T(t)$ is used with equation 4.1 to stop the heating cycle at time t_h such that

$$F_o(t_h) \geqslant F_o^d \qquad (4.6)$$

This approach was used by Mulvaney and Rizvy (1983), who used a thermocouple-fitted test can in every retort batch to obtain actual $T(t)$ readings. The system proposed by Navankasattusas and Lund (1978) was also planned to measure can center temperature on-line. In commercial practice these methods are cost-prohibitive with regard to production efficiency and would be viewed as impractical. Instead of measuring $T(t)$ at the can center, $T(t)$ can be predicted for arbitrary variations in $T_R(t)$, if the function f in equation 4.3 is known. In general, analytical forms of f are of limited use for arbitrary variations of $T_R(t)$. Teixeira and Manson (1982) described a numerical finite difference approximation to equation 4.4 for predicting $T(t)$ from $T_R(t)$ (the development of the mathematical model based on this approximation is described in Chapter 5). With a finite difference model $T(t)$ can be predicted for truly arbitrary variations in $T_R(t)$, and of course use of a model eliminates the need for having a test can with an actual thermocouple inserted in it. $F_o(t)$ can be monitored in real time by using $T(t)$ calculated from the finite difference approximation to equation 4.4 with measured boundary (retort) temperature $T_R(t)$. The heating process could continue until a time t_h such that

$$F_o(t_h) \geqslant F_o^d$$

thus satisfying the required sterilization F_o^d for arbitrary variations in retort temperature $T_R(t)$.

However, the contribution to F_o from cooling [the second integral in equation (4.1)] should not be neglected for a conduction heating product. Depending on the can center temperature at the start of cooling, the size of the can, and other factors in a conducting heating situation, the contribution to F_o from cooling could be as much as 40 percent or greater. This situation is unlike that with convection-heated products, where the container is agitated and therefore the product is rapidly cooled, making a small contribution to F_o during cooling. Therefore to avoid gross overprocessing of conduction-heated product, the cooling F_o must be considered. Also, as mentioned earlier, there is no way to control the possible contribution to F_o from cooling once cooling is under way.

Thus, while the contribution to sterilization (lethality) from cooling cannot be neglected, it is not a constant and cannot be controlled. Therefore, it can only be estimated before cooling is actually started, which can be done through simulation of the cooling cycle assuming a constant cooling water temperature in the retort. On a Digital Equipment Corp. (DEC) PDP 11/23 computer, the computation time required to simulate the cooling process was 9 s. Since only 1.5 s was required to calculate center temperature and F_o, the time interval between retort temperature readings could be as little as 12 s.

System Performance

A schematic diagram of a simple computer-based retort control system is shown in figure 4.1. As with any control system, the retort is provided with sensors to measure temperature, pressure, and water level along with various valves and switches to control the flow of air, water, and steam into and out of the retort. The controller in the system is a small digital computer, or microprocessor, which can be programmed to execute a specific sequence of control functions, as well as to read and interpret input signals, make decisions, and send output signals to the system. The microprocessor also reads input data, takes instructions, and prints output data through the keyboard printer, which serves as the operator's means of access and communication with the system. Figure 4.1 also shows an interface module between the retort and the microprocessor. This module is required to convert the analog signals received from the sensing instruments in the retort into electronic digital signals required by the microprocessor, as well as to convert digital output signals from the microprocessor back into analog and contact signals to actuate the valves and switches at the retort.

Although only one retort is shown in the figure, several retorts can be controlled in sequence with one computer and interface module, depending on the type of computer selected for the system. In this way, one system can control an entire cookroom operation, keeping track of the various

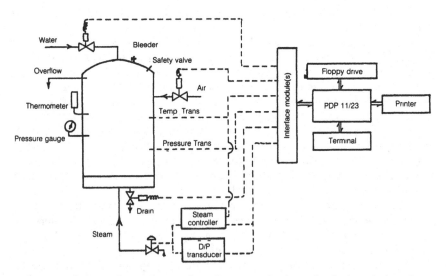

FIG. 4.1. Schematic representation of a computer-based retort control system.

steps in the operating cycle of each retort, including venting, cooking, and cooling. The control functions in a typical system would include: a security check of all input data to be sure that the correct process has been selected for the product to be retorted; rigorous control of the venting cycle, retort temperature, and process time during the cooking and cooling cycle; and printed documentation of all input data and times and temperatures of the operating cycles for each retort load, to provide a complete set of batch records in compliance with FDA record-keeping requirements.

The control logic diagram for the retort thermal process is given in figure 4.2. The input data are first checked against specifications for the product. If they agree, steam is turned on, and the computer completes the venting cycle. The retort temperature rises, and the computer, through a controller interface, attempts to maintain the design retort temperature T_R^d. As the heating cycle continues, retort temperature $T_R(t)$ is read at intervals of time ($\triangle t$), and $T(t)$ is then calculated from $T_R(t)$ by using the process simulation model with the finite difference approximation of equation 4.3. By using equation 4.1, $F_o(t)$ is then calculated. The input data includes a specified F_o^{heat} value, which is the F_o value normally achieved at the end of heating for the design total F_o value of F_o^d. When $F_o(t)$ exceeds F_o^{heat}, the computer also simulates the cooling cycle in addition to calculating $T(t)$ and $F_o(t)$. If the $F_o(t)$ accumulated so far, together with the simulated contribution from cooling, exceeds the design total F_o value for the process F_o^d, that is, when

$$F_o(t) + (F_o^{cool}) \text{ simulated} \geqslant F_o^d$$

is satisfied, the computer turns off the steam and admits cooling water. The computer still keeps reading the retort temperature $T_R(t)$ and calculating $T(t)$ and $F_o(t)$. When $T_{center}(t)$, the calculated temperature at the can center, is below a certain specified value, cooling is ended by stopping the cooling water flow, and the water is drained prior to unloading the retort.

At the end of the process, therefore, complete documentation of measured retort temperature history $T_R(t)$, calculated can center temperature history $T(t)$, and the accomplished $F_o(t)$ is on file and can be made available in both tabular and graphical form through printer and plotter.

The control algorithm presented as a flow chart in figure 4.2, when implemented in real time with a computer, can ensure designed sterilization F_o^d in a product for arbitrary variations in retort temperature. The performance of this system for correcting deviations has been demonstrated for three different processes: a reference process with no deviation (fig. 4.3); a step drop and later step rise in process temperature (fig. 4.4); and a gradual drop and later step rise in temperature (fig. 4.5). The adjusted process times and the accumulated F_o values for the three process conditions are presented in Table 4.1.

For the reference process without any deviation (fig. 4.3), the control logic keeps the process time and the accumulated F_o close to their design values.

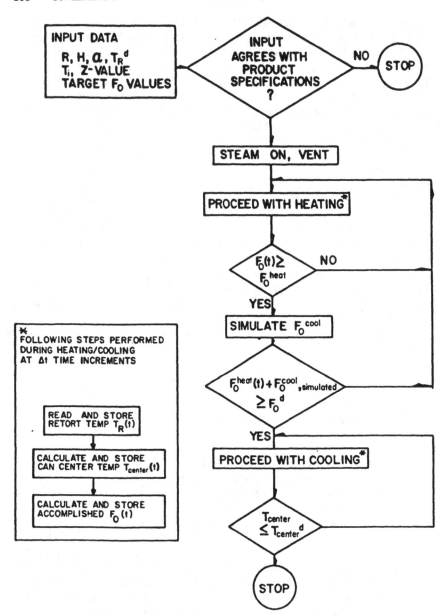

FIG. 4.2.   Flow diagram for computer control of retort operations with on-line correction of process deviations. (From Datta et al., 1986.)

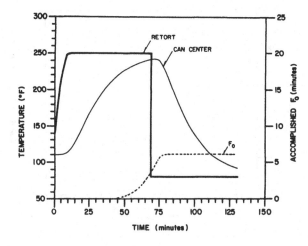

FIG. 4.3. Computer plot of reference process (no deviations) showing 66.8 minutes of heating time and an accomplished F_0 of 6.24. (From Datta et al., 1986.)

Next, a process with an idealized step-functional drop, shown in figure 4.4, was considered. The retort temperature undergoes an idealized step drop of 15°F for a period of 10 min from the fiftieth to the sixtieth min. The control logic extended the heating time to 72.2 min, resulting in a total accumulated F_0 value of 6.11, close to the required total F_0 of 6.0.

A process that experienced a gradual drop after about 40 min into the cook cycle was considered next (fig. 4.5). The drop in temperature continued for 20 min, with a rapid return to set point. The control logic extended the heating time to 74.6 min and still maintained an F_0 of 6.18.

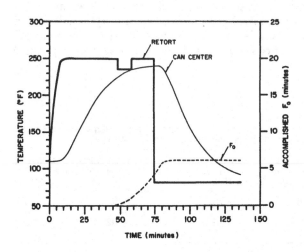

FIG. 4.4. Computer plot of a process that experienced a step functional drop of 15°F in temperature between the fiftieth and the sixtieth minute and still maintained an F_0 of 6.02. (From Datta et al., 1986.)

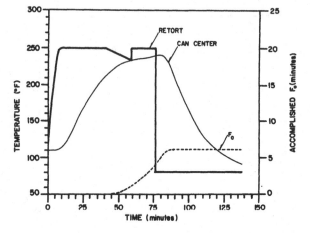

FIG. 4.5. Computer plot of a process that experienced a linear (0° to 20°F) drop in temperature between the fortieth and the sixtieth minute and still maintained an F_o of 6.18. (From Datta et al., 1986.)

The step-functional drop and the gradual drop in temperature are just two examples of process deviations in which the on-line control system is able to maintain the desired sterilization. The on-line control system predicts the can center temperature by using a finite difference analog of equation 4.4. It then uses this temperature value in a numerical (trapezoidal) method of integrating equation 4.1 (the general method). Both these steps place no restriction whatsoever on the retort temperature variation. This enables the on-line control system to maintain the desired sterilization for arbitrary variations in retort temperature.

Other benefit areas include real-time operational information and complete documentation. Figures 4.3 to 4.5 show a complete history of the process, including retort temperature, calculated can center temperature, and calculated F_o. In addition to providing all the required documentation for regulatory and quality control compliance, these graphs will provide visual descriptions of the complete process history. The graphs can clearly show if the process went through any deviation and how the required F_o was still maintained.

TABLE 4.1. Adjusted Heating Times and Resulting Lethality (F_o) in Response to Process Deviations Using Proposed On-line Control Logic (From Datta et al., 1986.)

	Reference Process (Fig. 4.3)	Step Drop (Fig. 4.4) Drop Duration 10 min Drop Amount 15°F		Gradual Drop (Fig. 4.5) Drop Duration 20 min Maximum Drop 20°F	
Heating time (min)	66.8	72.2		72.2	74.6
Total F_o (min)	6.24	6.11		6.02	6.18

ULTRAHIGH TEMPERATURE PROCESS CONTROL
FOR ASEPTIC SYSTEMS

Recent advances in aseptic packaging technology have sparked a growing new interest in this method of food preservation throughout the food industry (Ellis, 1982; Wagner, 1982). Basically, any food product that can be pumped through a heat exchanger can potentially benefit from aseptic processing and packaging. In these systems the product is thermally processed by being quickly heated, held, and cooled through a series of heat exchangers and a holding tube prior to aseptic filling into presterilized containers. Continuous sterilization under such ultrahigh temperature (UHT)/short time conditions can often result in improved product quality over conventional retort sterilization systems in which the product is sterilized in its container after filling and sealing (Simpson and Williams, 1974). Aseptic systems also provide an opportunity for filling very large containers for long-term bulk storage or shipment of processed or semifinished food products (Kafedshiev and Kolev, 1980; Nelson et al., 1974).

In designing the holding tube for continuous sterilization processes, it is sufficient for process evaluation to consider only the lethal effect achieved while the product is resident in the holding tube. This simplifies the heat transfer model by allowing the assumption of isothermal conditions in an insulated holding tube. Because of laminar flow, the resulting parabolic velocity distribution imposes a residence time distribution, that is, all parts of the product do not remain in the holding tube for the same length of time (fig. 4.6). Normally, the process requirements are determined by calculating the fastest particle velocity (hence, minimum holding time) from viscometric data. More often, however, processors are encouraged to make a worst-case assumption that their product behaves as a Newtonian fluid

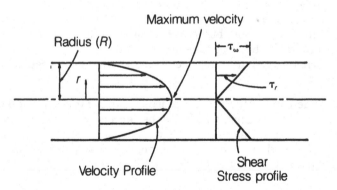

FIG. 4.6.   Velocity and shear-stress profile for laminar fluid flow in a holding tube with a circular cross section, where τ_w is shear stress at tube wall and τ_r is shear stress at any radial distance r from center streamline. (From Rao and Anatheswaran, 1982.)

and to choose a maximum velocity that is twice the bulk average velocity and thus provide a safety factor in accordance with good manufacturing practice (Palmer and Jones, 1976). The lethal effects on those portions of the product having a residence time greater than the fastest particle are not considered, which provides an additional safety factor.

In this way process evaluation is based on single-point lethality (fastest-moving particle) and is analogous to evaluation of a retort sterilization process on the basis of lethality F_o at the can center. Charm (1966) first suggested the concept of integrating lethality across the holding tube to account for the different residence times of various streamlines from the centerline to the tube wall; he later revised the mathematical expression for this integral (Charm, 1971) on the basis of comments by Beverloo (1967). This concept is analogous to Stumbo's (1965) integrated lethality F_s across a can of conduction-heated food from can center to can wall.

Our case study example describes the use of a computer model to examine both single-point and integrated lethality in an insulated holding tube for products with flow behavior characteristics ranging from pure non-Newtonian pseudoplastic to dilatant properties. It also describes a concept for on-line measurement and control of the sterilization process to accommodate batch-to-batch variations in product viscosity and other possible contributions to process deviations.

Description of the Model

The computer model was developed under the following assumptions:

1. Fluid flow is essentially isothermal in the insulated holding tube.
2. The fluid is characterized by the power-law model with no critical yield stress, and can exhibit flow behavior properties ranging from pseudoplastic through Newtonian to dilatant.
3. The product is homogeneous.
4. Lethal effects contributed during product heating and cooling before and after the holding tube are ignored.
5. Thermal inactivation of bacterial spores follows first-order reaction kinetics (logarithmic survivor curves).

For single-point lethality the process is evaluated on the basis of the minimum residence time, which is experienced by the fastest-moving particle at the center streamline. This maximum velocity V_{max} can be determined as a function of the bulk average fluid velocity $\overline{V^2}$ and the flow-behavior index s of the fluid (Charm, 1971):

$$V_{max} = \overline{V} \frac{(3s + 1)}{(s + 1)} \qquad (4.7)$$

Charm (1971) also suggested an expression for the integrated lethality across the various streamlines in a holding tube:

$$N = N_o \int_o^R 10^{-\theta r/Dt} \, 2\pi V_r dr \qquad (4.8)$$

where N = number of surviving spores emerging from the holding
             tube
       N_o = initial concentration of spores per unit volume
       D_t = time for the logarithmic survivor curve to pass through
             one logarithmic cycle at the holding tube temperature
       θ_r = residence time of a streamline in the holding tube =
             L/V_r, where L = length of holding tube
       r = radial distance to a given streamline
       R = radius of holding tube
       V_r = velocity of a given streamline

To carry out the integration in equation 4.8, Teixeira and Manson (1983) showed how the velocity profile can be expressed as a function of the volumetric flow rate, flow behavior index, radius, and radial distance:

$$V_r = \frac{Q}{\pi R^2} \left[1 - \left(\frac{r}{R} \right)^{\frac{s+1}{s}} \right] \frac{3s + 1}{s + 1} \qquad (4.9)$$

The model was completed by inserting equation 4.9 for V_r in equation 4.8 and expressing θ_r by $\theta_r = L/V_r$ to calculate the integrated flow rate of surviving spores emerging from all holding tube streamlines of length L. Integration was performed numerically by computer iteration across streamlines from the center line ($r = 0$) to the holding-tube wall ($r = R$).

Integrated and Single-Point Lethality

Single-point (F_o) and integrated (F_s) lethalities can be calculated for a variety of holding tube conditions, flow behavior properties, and thermal inactivation kinetics.

To study the effect of flow behavior index and decimal reduction time on process lethality, the model was used to calculate both F_o and F_s over a range of s and D values. A reference process was selected on the basis of a Newtonian fluid design ($s = 1$) with $F_o = 7$ min to establish a minimum holding tube length for a given set of processing conditions (tube diameter, temperature, and volumetric flow rate). Under this same set of conditions F_s was calculated with three different values of decimal reduction time at 250°F ($D = 0.5$, 1.0, and 1.5 min) to represent different microorganisms of concern in the reference fluid. This work was then repeated with different values of s to represent non-Newtonian fluids that could exhibit either pseudoplastic ($0 < s < 1$) or dilatant ($1 < s$) flow behavior properties.

The results of this work are described by the family of curves in figure 4.7, which shows how both F_o and F_s vary as a function of s for a fixed holding tube design and how the difference between F_o and F_s depends on both s and D. The difference between F_o and F_s diminishes as s approaches

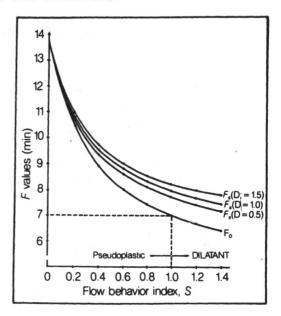

FIG. 4.7. Single-point (F_o) and integrated (F_s) holding tube lethality versus flow behavior index s in both pseudoplastic and dilatant ranges for integrated lethalities based on different heat resistances (D) of bacterial spores. (From Teixeira and Manson, 1983.)

zero. Such flow behavior would be characterized by a much less pronounced velocity profile, in which the residence-time distribution is more nearly uniform across the holding tube streamlines. These results also show that for most pseudoplastic foods ($s < 0.5$), the difference between F_x and F_o is less significant than the difference between the actual lethality achieved (either F_o or F_x) and the F_o designed on the basis of Newtonian fluid assumptions ($s = 1$). This conclusion in turn suggests that process design for such foods could ignore consideration of integrated lethality across the holding tube but should be based on the maximum velocity (hence minimum residence time) dictated by the actual flow behavior index of the fluid (eq. 4.7).

The findings further illustrate how the popular industry practice of process design based on Newtonian fluid assumptions ($s = 1$) can often be overly conservative for pseudoplastic foods. For example, a fluid with $s = 0.4$ (not uncommon) would receive an actual F_o value approximately 30 percent greater than the design F_o based on $s = 1$. In addition, the total integrated lethality could contribute a further safety factor on the order of 10–15 percent, depending upon the organism of concern. Just as with conduction-heated canned foods, integrated lethality becomes increasingly more significant than single-point or center-point lethality in continuous UHT sterilization processes as the heat resistance or decimal reduction time of the spoilage organism increases, as with thermophilic bacterial spores.

On-Line Control Concept

The FDA's low-acid canned food regulations governing aseptic processing and packaging systems specify, in part, that records must be maintained showing adequate control of product temperature in the holding tube and final heater outlet, as well as of product flow rate and heating medium flow rate. The regulations further specify that when a process deviation occurs, as when temperature in any part of the process is less than in the scheduled process, or when critical factors are out of control for any product or container system, the portion of product so affected must be either fully reprocessed or set aside for further evaluation by a competent processing authority. Thus, good process control of continuous UHT sterilization systems is of critical importance to the food processor.

The work presented thus far shows how control of temperature and flow rate in the holding tube can only be meaningful if the flow behavior properties of the fluid food product remain unchanged. This is particularly true of the fluid viscosity, which will affect the flow behavior index upon which the scheduled sterilization process design was based.

In a typical production operation the fluid food product is normally fed to the continuous UHT sterilization system from batch tanks in which the raw ingredients are mixed and blended together. Often, these batch tanks hold only enough product to feed the filling line over a limited period of operation. Several batches may be prepared in sequence throughout the day, while the product feed is frequently switched from one batch tank to another to maintain continuous feed. Batch-to-batch variations in ingredient compositions and mixing and blending conditions could result in significant changes in the flow behavior properties of the fluid. Subsequent shear stress from downstream pumps and valves acting on a different set of viscometric properties would change the flow behavior index of the fluid and hence the velocity profile.

In their discussion on the viscometric behavior of power-law fluids, Manson and Cullen (1974) showed how the flow behavior index and viscometric constant K of a pseudoplastic fluid can be determined from a double logarithmic plot of apparent viscosity versus shear rate (at a given temperature), which yields a straight line having a negative slope equal to s - 1. The value of K is determined by the intercept at a shear rate of $1s$ - 1. Figure 4.8 illustrates the viscosity versus shear rate data for a typical pseudoplastic food product and shows how temperature variation has no significant effect on the flow behavior index over the range of temperatures of interest to food sterilization. The mathematical expression for this relationship is given by Manson and Cullen (1974) as:

$$\mu = K\gamma^{s-1} \qquad (4.10)$$

where μ = apparent viscosity
          K = viscometric constant
          γ = shear rate

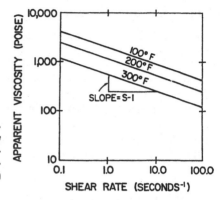

FIG 4.8. Viscosity curve for a pseudo-plastic non-Newtonian fluid showing how the flow behavior index s can be determined from apparent viscosity measurements at two different shear rates. (From Manson and Cullen, 1974.)

This relationship suggests that in-line monitoring of the flow behavior index could be accomplished indirectly from two apparent viscosity readings at two different shear rates. One method is to have two in-line viscometers operating at different shear rates. Also, Rao and Bourne (1977) showed that the plastometer of Eolkin (1957), although designed to measure consistency, could provide a continuous signal that is in effect "proportional to the difference in apparent viscosities at two different shear rates." Eolkin's plastometer is a relatively inexpensive instrument, made up of a simple flow bridge network, in which the pressure difference measured between two points is proportional to the difference in apparent viscosities at two different shear rates. A schematic diagram of Eolkin's plastometer is shown in figure 4.9.

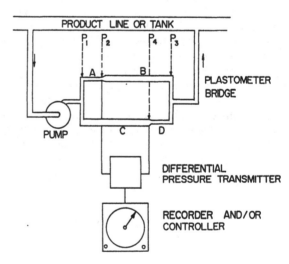

FIG. 4.9. Eolkin plastometer for continuous measurement of apparent viscosity at two different shear rates in fluid foods. (From Rao and Bourne, 1977.)

One of the above instrument systems could be installed ahead of the holding tube and calibrated to measure apparent viscosity difference for a given product. This would serve as an in-line monitor to ensure that variations in product viscosity do not alter the velocity profile in any way that would prevent maintenance of minimum residence time. Such a device could also be used to trigger a divert valve if the measured signal indicates that minimum residence time cannot be maintained. The ultimate application for such instruments, however, would be in real-time on-line control of the process, which could be achieved by placing the monitoring instrumentation further upstream to allow some time to elapse as the product fluid flows from the monitoring point before entering the UHT sterilization system. Care would need to be taken that there are no process operations between these points that would alter the flow behavior of the fluid. The instruments, if so placed, could operate as part of a control system that would compensate for any change in velocity profile by adjusting holding tube temperature, if necessary, and would avoid the product loss and downtime caused by a product-divert situation, with subsequent resterilization and start-up of equipment.

The above system would likely make use of a small industrial microcomputer (microprocessor), which would receive signals from the viscometers or plastometer via an ADC. The microprocessor could be programmed to make the necessary temperature correction calculation in the event of any change in fluid flow properties and to send the appropriate control signal to the UHT heat exchanger to raise the product temperature before the affected product reaches the holding tube.

MULTIEFFECT EVAPORATION IN JUICE CONCENTRATION

This case study example describes an application of on-line computer control in the operation of commercial multieffect evaporators for reducing energy consumption in the manufacture of frozen concentrated orange juice (FCOJ) in the Florida citrus industry. Evaporation is the most energy-intensive unit operation in the manufacture of this product and accounts for nearly 50 percent of the total energy used by the U.S. citrus processing industry (Chen et al., 1981). Most evaporators used in making concentrate in the Florida citrus industry are of the single-pass, falling-film, long-tube type using high temperature–short time principles and known as *thermally accelerated short-time* (TASTE) evaporators (Cook, 1963). The TASTE evaporator consists of a series of units arranged in order of descending pressure (and therefore descending boiling point) so that the vapor from any one unit (effect) heats the unit of next lower pressure. The last effect usually consists of two to three stages.

To secure proper evaporation in each effect, two important requirements must be met: first, the necessary heat of vaporization must be supplied to the liquid, and second, the vapor evolved must be separated from the liquid and not allowed to accumulate. Each effect, therefore, must have its proper heat transfer area, vapor liquid separator, air vents, condensate removal, and so on. These requirements underline the importance of proper design, both thermally and mechanically, and also demand proper maintenance of the system. Failure of either proper design or maintenance will result in lower efficiency or loss of capacity (Standiford, 1963).

As a first approximation in calculating energy requirements for an N-effect evaporator, one-Nth of the water removed occurs in each effect; and since process steam is provided for the first effect, only one-Nth pound of steam is required to evaporate each pound of water. The actual energy requirement is best predicted by energy and mass balance analyses. The data on commercial TASTE evaporators is shown in figure 4.10. Efficiency is often expressed as *steam economy,* that is, the ratio of pounds of water evaporated to pounds of steam used. The steam economy should be roughly equal to 0.85 N for the TASTE evaporator (Rebeck, 1976); but, calculations from field data in figure 4.10 show that only 0.6 N to 0.82 N has been observed in practice. The shaded area is the region in which potential energy savings could be obtained through improvement of operation or design or addition of effects.

On-Line Computer Control System

The traditional control system for most TASTE evaporators consists of two pneumatic control valves, one to control steam flow and the other to

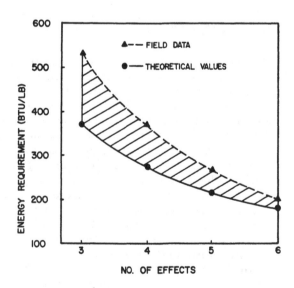

FIG. 4.10. Energy requirement for multiple-effect evaporators. (From Chen et al., 1981.)

control feed juice flow. Under manual operation the steam flow is usually set at constant pressure and the feed flow is adjusted manually according to the desired pumpout Brix value (juice concentration). The pressure of the vapor leaving the last effect is determined by the cooling water. When the system is in operation, a distribution of temperature among effects is established, and a steady operation is reached after a certain elapsed time. This steady-state condition to produce the desired pumpout Brix is the target of the evaporator operators, and how fast this desired condition can be reached depends largely on their experience and skill, but how long it can be maintained is beyond their control. Desired pumpout Brix can increase or decrease several degrees within a few minutes without changing any operating settings, but any attempt to correct the deviation may require several changes of feed rate and/or steam rate by trial-and-error methods. Prolonged deviation from the desired value may result not only in an off-specification product but also in loss of useful energy.

The conventional control system consists of two feedback control loops for steam and feed flow. A diagram of a generalized feedback control loop is shown in figure 4.11. The requirement of frequent changes in control operation due to natural fluctuations in variables dictates a need for automation. On the basis of these considerations, a process control microcomputer with 16-bit processing and 12-bit data acquisition capability (MACSYM II, Analog Devices, Westwood, MA) was selected as a stand-alone, interactive measurement and control system.

A six-effect, eight-stage TASTE evaporator, designed for a 20,000-lb/h evaporation capacity, was first instrumented for automation and is schematically shown in figure 4.12. Steam condensate temperature in the first effect was used for steam control, and the fifth stage (fourth effect) juice temperature was used for feed flow control. The steam control was achieved

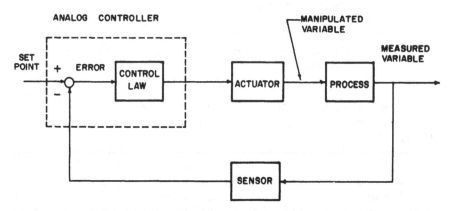

FIG. 4.11.   Basic conventional feedback control loop. (From Chen et al., 1981.)

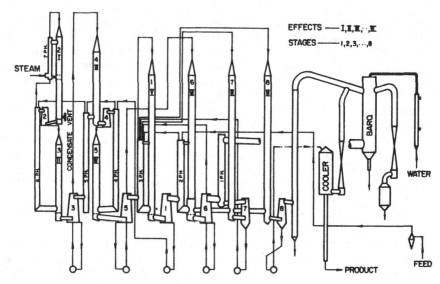

FIG. 4.12. A six-effect eight-stage TASTE evaporator used in the Florida citrus industry. (From Chen et al., 1981.)

by the following proportional-integral-differential (PID) control model (Smith, 1970):

$$V_n = K_c E_n + (K_c T / T_i) \sum_{i=0}^{n} E_i + (T_d K_c / T) (E_n - E_{n-1}) + V_r \quad (4.11)$$

where K_c = proportional gain
         T_i = reset or integral time
         T_d = derivative time
         T = sampling time
         E_i = value of the error at the i-th sampling instant
                $(i = 0, 1, \ldots, n)$
         V_n = value of manipulated variable at the n-th sampling instant
         V_r = reference value at which the control action is initialized

Feed flow was sufficiently controlled by a proportional-integral (PI) model that does not contain the third term on the right-hand side of equation (4.11).

A schematic diagram for conversion of manual to digital computer control is shown in figure 4.13 where A/I stands for analog input, A/O stands for analog output, M_n is a measured variable, and R_n is a reference value (set point). A schematic diagram of the control scheme is shown in figure 4.14. Both steam flow and feed flow are manipulated variables. Temperature measurements are shown in figure 4.15. The measured steam condensate

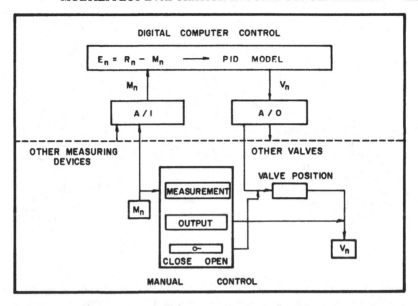

FIG. 4.13.   Schematic diagram for converting manual to digital computer control. (From Chen et al., 1981.)

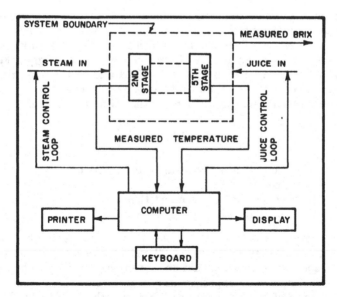

FIG. 4.14.   Schematic diagram of on-line computer control scheme for a multiple-effect evaporator. (From Chen et al., 1981.)

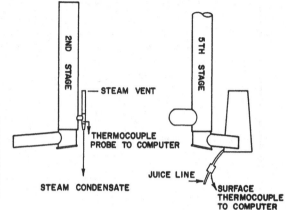

FIG. 4.15. Thermocouple locations for measuring steam and product temperatures and control variables in computer control of multiple-effect evaporators. (From Chen et al., 1981.)

temperature (an intermediate variable) was fed back to control inlet steam, forming one loop, and the measured fifth-stage juice temperature was fed back to control feed flow, forming the other loop. The measured eighth-stage juice concentration (Brix) was the controlled variable.

System Performance

Experiments were conducted to demonstrate the energy savings achievable by automatic control. Heating steam rates were measured by piping the steam condensate from the first effect to a collection tank. The steam condensate was collected at 10- to 15-min intervals and weighed. Feed rates were measured by a flowmeter, and feed Brix and final-stage Brix degrees were measured at 10- to 15-min intervals by refractometer.

The evaporation rates were then calculated by the formula

$$E = W \left(\frac{C - F}{C} \right) \tag{4.12}$$

where E = pounds of water evaporated per hour
         F = feed juice Brix
         C = concentrate or product Brix
         W = pounds of feed juice per hour

The evaporator was found to operate at 70–95 percent of the design capacity. Steam economy ratios were 4.63 and 4.94 for manual and automatic control, respectively, showing that energy savings achievable by automatic control techniques were approximately 6.7 percent.

On the basis of the data obtained, along with the other field data on commercial TASTE evaporators (Chen et al., 1979), it was possible to evaluate the economic feasibility of automatic control. Reduction of energy

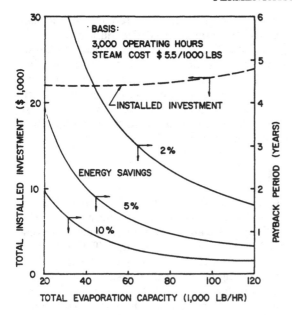

FIG. 4.16. Investment and payback period for computer control of a four-effect evaporator. (From Chen et al., 1980.)

consumption depends on several factors, such as evaporation capacity, number of effects, and operation efficiency. Cost savings depend upon computer cost against actual fuel savings and fuel cost. Estimates of investment and payback period for automatic control of a four-effect evaporator are shown in figure 4.16 (Chen et al., 1980). The total investment cost increased very little with increasing total evaporation capacity, which suggests that it would be advantageous to use one computer to control several units. Steam and juice flows were controlled in the experiments, but it was found that steam flow control alone was acceptable in routine operations.

FERMENTATION PROCESS CONTROL

The goal of computer control in fermentation operations is to achieve a prespecified and reproducible fermentation pattern, optimized with respect to the stoichiometry and kinetics of the substrate-to-product conversion and with respect to energy utilization and volumetric productivity. This goal requires a high degree of process simulation, along with computerized process control (Cooney, 1979).

A great number of kinetic models for batch, semicontinuous, and continuous fermentations in aerobic and anaerobic, liquid and solid substrates has been published (Bosujak et al., 1979; Fawzay and Hinton, 1980; Garcia and Grillione, 1982; Meiering et al., 1978; Okuda et al., 1981; Rolf and Lim, 1982; and Weigand et al., 1979). The models relevant to ethanol fermentation were generally based on the Michaelis-Menten and Monod theories of microheterogeneous reaction systems and differ mainly in the description of product and substrate inhibition and in the statistical treatment of parameter evaluations. Batch fermentation models for wine production have been published by Boulton (1980), and elements from these concepts were combined with a more detailed description of microbial growth kinetics, carbon dioxide and ethanol production, and heat transfer in the fermentation model used in this case study. A detailed description of this model can be found in Meiering and Subden (1983).

Process System Description

In this case study example, fermentation experiments were performed with a substrate (must) prepared from a grape juice concentrate of the French hybrid variety Elvira. Four 100-L stainless steel fermentation vessels, as illustrated in figure 4.17, were used in the study. The must was inoculated with the Geisenheim strain of *Saccharomyces cerevisiae.* The control system provided for monitoring of substrate density, temperature, and carbon dioxide production and for control of yeast distribution and substrate temperature, all functions performed by a Rockwell AIM65 mi-

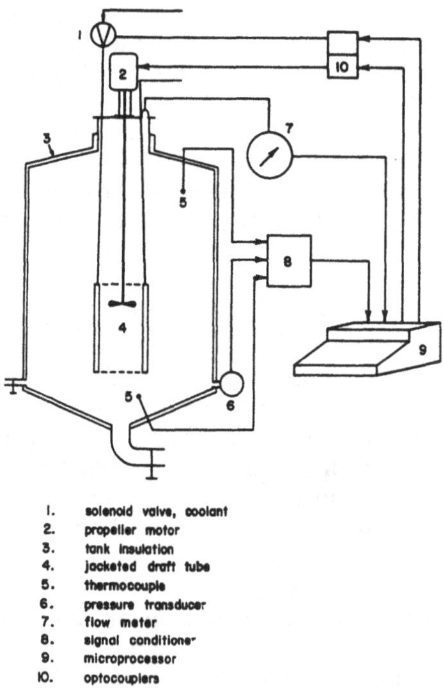

FIG. 4.17. Fermentation control scheme. Shown on the diagram are the: solenoid valve for coolant (1); propeller motor (2); tank insulation (3); jacketed draft tube (4); thermocouple (5); pressure transducer (6); flowmeter (7); signal conditioner (8); microprocessor (9); and optocouplers (10). (From Meiering and Subden, 1980.)

1.	solenoid valve, coolant
2.	propeller motor
3.	tank insulation
4.	jacketed draft tube
5.	thermocouple
6.	pressure transducer
7.	flow meter
8.	signal conditioner
9.	microprocessor
10.	optocouplers

croprocessor. Copper-constantan thermocouples, a positive displacement gas flowmeter producing a 5-V pulse upon completion of each revolution, and a Schaevitz P 500 high-accuracy pressure transducer were used with Analog Devices hardware components, as shown in figure 4.18.

Batch fermentations in 100 L vessels, as shown in figure 4.17, were monitored for the first 200 h after inoculation. At a relative height $h = 1.5\ d$ (where d is diameter) the surface area of the tank was 1.21 m². The tanks had insulated walls, and one had a jacketed draft tube with a total surface area of 0.28 m². Water was circulated through it as coolant at an hourly flow rate of half the tank volume. The convective heat transfer coefficients of the tank surface and the draft tube surface were estimated.

Yeast distribution throughout the must was enhanced by the mixer blades positioned in the center of the draft tube. During the period of intensive microbial activity, the mixer, together with the coolant circulation, was activated by the microprocessor. The thermocouple scanning interval was set at 30 s to keep a uniform substrate temperature of 14°C, with fluctuation and gradients between top and bottom layers not exceeding 0.3°C. Temperature differences between 2 and 5°C had been observed to occur in an unstirred substrate. The must density was scanned in 30-min intervals, and carbon dioxide flow rates were recorded.

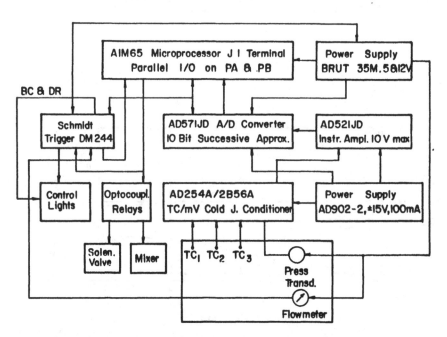

FIG. 4.18.   Flow chart of signal processing for fermentation control scheme. (From Meiering and Subden, 1980.)

System Performance

Good agreement was reached between measured and simulated batch process data on substrate density, carbon dioxide production, and cell concentration as shown by the results of a typical experiment (fig. 4.19). No continuous flow experiments were performed, but the simulated data also agreed well with experimental data published by Cahill et al. (1980) (fig. 4.20). A higher ethanol tolerance and maximal growth rate, which is typical for yeasts used in champagne production, was assumed. Also, a lower saturation constant, which increased with the third power of the dilution rate,

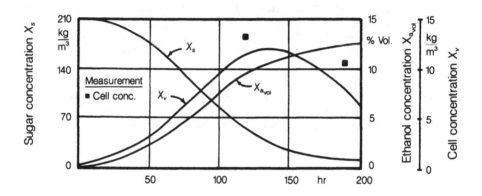

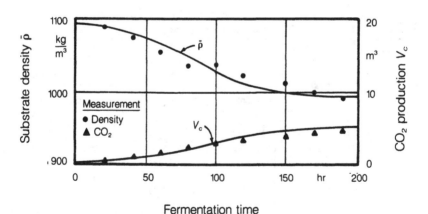

Fermentation time

FIG. 4.19.   Batch fermentation of must with *Saccharomyces cerevisiae*. Initial sugar concentration S_0 = 210 kg/m³. Initial yeast cell concentration A_0 = 0.5 kg/m³. (From Meiering and Subden, 1980.)

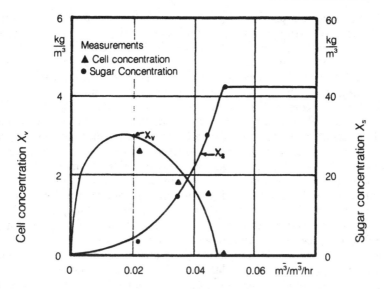

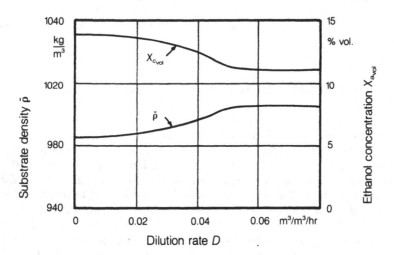

FIG. 4.20.   Continuous flow fermentation of wine. S_0 = 42.7 kg/m³, A_0 = 11.3 percent by volume. (From Meiering and Subden, 1980.)

was observed in a preliminary evaluation. Process optimization could be achieved by varying parameters and input conditions, and selected results can be stored as decision criteria before the fermentation begins. The ripple effects of even minor changes in operating conditions on the entire process could be clearly recognized and evaluated from the simulations.

Cooling requirements were based on an average temperature calculated by the microprocessor. The mixer activation could be programmed to stop after the completion of the primary fermentation period to allow settling of the suspended solids and yeast cells. An accurate measurement of carbon dioxide production may be too complicated or expensive in a commercial winery owing to the high requirements on leakage control. Density and temperature measurements, however, appear technically and economically feasible and will allow effective monitoring, control, and inventory of a production unit in terms of quality criteria, process time, and yields. Additional sensors—for example, for enzyme activity, redox potential, and cell fluorescence—could improve the resolution of process dynamics, and extended software could tie quality criteria such as concentrations of flavor and aroma compounds more definitely to the kinetic pattern.

In summary, this case study example has shown that the reaction kinetics of batch and continuous-flow ethanol fermentation in must can be described in simulation models suitable for interactive, on-line computer process control. It has also shown that must temperature and density can be readily measured with conventional corrosion-resistant sensors and can serve as criteria for basic fermentation control and, moreover, that a microprocessor memory of 8K bytes can store the algorithms of an elementary, on-line control system for several fermentation tanks.

COMPUTER CONTROL IN BIN DRYING OPERATIONS

Our final case study will describe a project involving on-line computer control in the drying of hops and should serve as an example of computer control applied to any bin drying operation. Hops are cone-shaped flowers produced on climbing vines of the hop plant, which are harvested and dried for use as an important flavoring ingredient in the malt beverage industry.

According to Thompson et al. (1982), hops are dried to a final moisture content of 10 percent from a moisture at harvest of 78 percent. In the United States the driers, or kilns, are heated forced-air systems. The hops are piled to a uniform depth of 1 m on a large floor and air that has been heated to 60–70°C (140–158°F) is forced through at a rate of 0.13–0.20 m/s (25–40 cfm/ft^2). Total drying time is 8 to 12 h; however, Henderson and Miller (1972) have suggested that drying time can be significantly reduced if the drying air temperature and flow rate are increased above normal levels at the beginning of drying. This method is called modified air flow and temperature drying (MAT). However, sustained high temperatures can affect hop quality, and temperature must therefore be reduced during the drying period. Air flow can be as high as 0.3 m/s (60 cfm/ft^2) at the beginning of drying when the hops are heavy, but as drying proceeds, the flow rate must be gradually reduced to prevent the hops from lifting off the floor of the kiln. Ordinary thermostatic controllers for temperature and manual control for air flow could be used to provide the conditions for MAT drying. However, microcomputers offer the potential for more precise control, with much less potential for operator error than existing hop kiln control systems. Microcomputers are also well suited for displaying kiln operating data, which is useful in determining when the hops are finished drying. There is no moisture meter that will determine average hop moisture content in the kiln.

The system hardware as described by Thompson et al. (1982a) is shown in figure 4.21. An Apple II microcomputer (with 48K bytes of random access

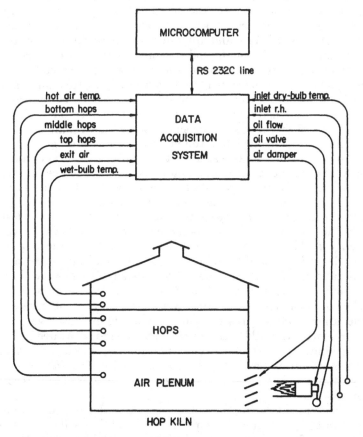

FIG. 4.21.   Schematic diagram of control equipment in hop drying system. (From Thompson et al., 1982a.)

memory) was selected for the project. This is a widely used microcomputer, for which a large amount of peripheral hardware and software has already been developed. The microcomputer was fitted with a video display screen, two 5¼ inch floppy disk drives, and an Epson MX 80111 dot matrix printer. Kiln operating data were transmitted to the microcomputer through an Analog Device μMAC 4000 data acquisition system, which converted thermocouple signals to Celsius temperature readings and voltage inputs from a humidity probe and oil flow sensor to equivalent digital readings. On command from the microcomputer, the data acquisition system sent data to the microcomputer over an RS-232C data communications link. The microcomputer controlled separate servo motors for a fuel oil flow valve and an air damper through a digital output port on the μMAC 4000.

Process Control Logic and System Performance

A block diagram of the software is shown in figure 4.22. The program began by asking the operator the depth of the hops, bulk density, initial moisture content, temperature set points and limit, and air flow limit. Then the temperature, oil flow, and humidity data were obtained from the data acquisition system. Dry-bulb and wet-bulb temperatures were converted to relative humidity; instantaneous oil flow was converted to cumulative oil use; air flow was estimated by using oil consumption and temperature rise of the heated air; and hop moisture was estimated by using initial moisture, bulk density, and bed depth data to determine the quantity of water removed by the drying air since the last data sample.

The data were displayed on the video screen according to the example in figure 4.23. The display was a mixture of high-resolution graphics used to draw the lines and low-resolution graphics used to display the letters and numbers. The two types of graphics were mixed on the same screen by using two machine language programs, both user-donated programs obtained through the microcomputer dealer. The display was designed to present all the pertinent data in an easily readable form. Temperature, humidity, and air flow data were displayed on a schematic diagram of the dryer; elapsed time and cumulative oil use were displayed as moving bar graphs; and moisture content was plotted versus time to allow the operator to estimate when drying was done. After displaying the data, the program checked data entries to see if they were within acceptable ranges. For example, an open circuit in a thermocouple would produce a very high temperature, and this would cause an error message to be printed on the screen and an alarm to be sounded. The program also checked for rises in air temperature or air flow rate above prescribed limits. Data were printed every 15 min and at the end of drying to provide a hard copy record of the dryer operation.

The current operating data were used to calculate any adjustments needed in the position of the air damper or oil flow valve on the burner. The set point for air flow rate was based on the estimated moisture content of the hops. Air flow rate started at 0.3 m/s (60 cfm/ft²) and was reduced

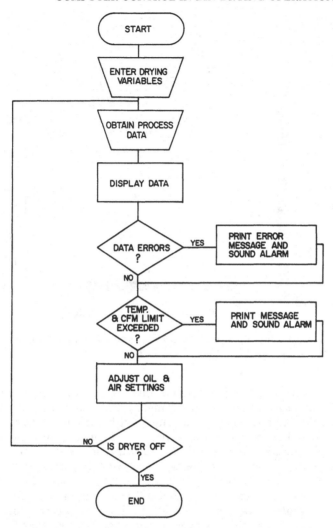

FIG. 4.22. Flow chart of on-line control logic for computer
control of hop drying process. (From Thompson et al., 1982a.)

gradually to 0.15 m/s (30 cfm/ft^2) as the hops dried. Air temperature was
set initially at 82°C (180°F) and dropped to a low temperature set point of
65°C (149°F) or 70°C (158°F) when the air temperature at a point 23 cm
(9 in) above the bottom of the hop bed reached the low-temperature set
point. This prevented the hops at or above the 23-cm (9-in) level from heating above the low-temperature set point.

The time and direction of servo motor rotation were determined by using

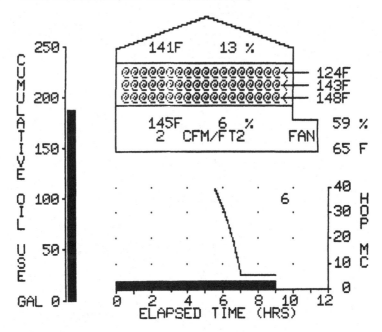

FIG. 4.23. Example of video display output in computer control of hop drying process. (From Thompson et al., 1982a.)

the velocity form of the proportional-integral control algorithm. Control constants were estimated by using motor speeds and measuring rises in air flow rate and temperature associated with various positions of the air damper and oil valve. Estimated levels were adjusted while the dryer was operating to obtain relatively fast and stable operating conditions. The proportional constant for oil flow was decreased proportionally as air flow rate decreased. The program was halted for 30 s to allow the dryer to stabilize, and the program was repeated. When the dryer was turned off, the program operation was stopped and the final video screen display was printed. With the use of this control system, total drying time for a typical bin dryer batch of hops could be reduced from 12 to 8 hr.

REFERENCES

BEVERLOO, W. A. 1967. Survival of microorganisms in continuous HTST processes—an error and additional observations. Food Technology 21: 964.

BOSUJAK, M., TOPOLOVEC, V., and JOHANIDES, T. 1979. Growth kinetics and antibiotic synthesis during repeated fed-batch culture of streptomycetes. Proceedings of the 2nd International Conference on Computer Applications in Fermentation Technology. Biotechnology and Bioengineering Symposium No. 9. John Wiley & Sons, New York. p. 155–165.

BOULTON, R. 1980. The prediction of fermentation behavior by a kinetic model. American Journal of Enology and Viticulture 31(1): 40–45.

CAHILL, J. T., CARROAD, P. A., and KUNKEE, R. E. 1980. Cultivation of yeast under carbon dioxide pressure for use in continuous sparkling wine production. American Journal of Enology and Viticulture 31(1): 46–52.

CHARM, S. E. 1966. On the margin of safety in canned foods. Food Technology 20(5): 97.

CHARM, S. E. 1971. Fundamentals of Food Engineering. 2nd Ed. AVI Publishing Co., Westport, CT.

CHEN, C. S., CARTER, R. D., and BUSLIG, B. S. 1979. Energy requrements for the TASTE citrus juice evaporator. In Fazzolare, R. A., and Smith, C. B., eds. Changing Energy Use Futures. Vol. 4, pp. 1841–48. Pergamon Press, Elmsford, NY

CHEN, C. S., CARTER, R. D., and DEIMLING, C. J. 1981. Microcomputer control of commercial citrus TASTE evaporators. Transactions of the 1981 Citrus Engineering Conference, Vol. 27. American Society of Mechanical Engineers, Lakeland, FL.

CHEN, C. S., CARTER, R. D., MILLER, W. M. and WHEATON, T. A. 1980. Energy performance of a HTST citrus evaporator under digital computer control. ASAE Paper No. 80-6028, American Society of Agricultural Engineers, St. Joseph, MI.

COOK, R. W. 1963. High temperature–short time evaporation. Transactions of the 1963 Citrus Engineering Conference. Vol. 9. American Society of Mechanical Engineers, Lakeland, FL.

COONEY, C. L. 1979. Computer applications in fermentation technology—a perspective. Proceedings of the 2nd International Conference on Computer Applications in Fermentation Technology. Biotechnology and Bioengineering Symposium No. 9. John Wiley & Sons, New York, pp. 1–11.

DATTA, A. K., TEIXEIRA, A. A., and MANSON, J. E. 1986. Computer-based retort control logic for on-line correction of process deviations. Journal of Food Science 51(2): 480–483, 507.

ELLIS, R. F. 1982. Aseptically packaged juice and milk taking hold on U.S. market. Food Processing 43(1): 96.

EOLKIN, D. 1957. The plastometer—a new development in continuous recording and controlling consistometers. Food Technology 11: 253.

FAWZEY, A. S., and HINTON, O. R. 1980. Microprocessor control of fermentation processes. Journal of Fermentation Technology 58(1): 61–67.

GARCIA, A., and GRILLIONE, P. 1982. Ethanol production characteristics for a respiratory-deficient mutant yeast strain. Transactions of the ASAE 25(5): 1396–1399.

GETCHELL, J. R. 1980. Computer process control of retorts. ASAE Paper No. 80-6511, American Society of Agricultural Engineers, St. Joseph, MI.

GIANNONI-SUCCAR, E. B., and HAYAKAWA, K. I. 1982. Correction factor of deviant thermal processes applied to packaged heat conduction food. Food Technology 47(2): 642–646.

HENDERSON, S. M., and MILLER, G. E., JR. 1972. Hop drying—unique problems and some solutions. Journal of Agricultural Engineering Research 17: 281–287.

KAFEDSHIEV, I., and KOLEV, O. 1980. Processing and bulk storage of semifinished fruits and vegetables. Food Technology 34(7): 56.

MANSON, J. E., and CULLEN, J. F. 1974. Thermal process simulation for aseptic processing of foods containing discrete particulate matter. Journal of Food Science 39: 1084–1089.

MANSON, J. E., STUMBO, C. R., and ZAHRADNIK, J. W. 1974. Evaluation of thermal processes for conduction-heating foods in pear-shaped containers. Journal of Food Science 39: 276.

MANSON, J. E., ZAHRADNIK, J. W., and STUMBO, C.R. 1970. Evaluation of lethality and nutrient retention of conduction-heating food in rectangular containers. Food Technology 24(11): 109–113.

MEIERING, A. G., AZI, F. A., and GREGORY, K. F. 1978. Microbial protein production from whey and cassava. Transactions of the ASAE 21(3): 586–593.

MEIERING, A. G., and SUBDEN, R. E. 1983. Fermentation control by micro computers. ASAE Paper No. 83-6534, American Society of Agricultural Engineers, St. Joseph, Mi.

MULVANEY, S. J., and RIZVY, S. H.   1984.   A microcomputer controller for retorts. Transactions of the ASAE 27(6): 1964–1969.
NAVANKASATTUSAS, S., and LUND, D. B.   1978.   Monitoring and controlling thermal processes by on-line measurement of accomplished lethality. Food Technology 43(3): 79–83.
NELSON, P. E., SULLIVAN, G. H., and HERON, J. R.   1974.   Aseptic processing and bulk storage of fruit products. In Proceedings of 4th International Congress of Food Science and Technology 4(5).
OKUDA, W., FUKUDA, H., and MORIKAWA, H.   1981.   Kinetic expressions of ethanol production rate and ethanol consumption rate in bakers yeast cultivation. Journal of Fermentation Technology 59(2): 103–109.
PALMER, J. A., and JONES, V. A.   1976.   Prediction of holding times for continuous thermal processing of power-law fluids. Journal of Food Science 41: 1233.
RAO, M. A., and ANANTHESWARAN, R. C.   1982.   Rheology of fluids in food processing. Food Technology 36(2): 116.
RAO, M. A., and BOURNE, M. C.   1977.   Analysis of the plastometer and correlation of Bostwick consistometer data. Journal of Food Science 42: 262.
REBECK, H.   1976.   Economics in evaporation. Proceedings of the 16th Annual Short Course for the Citrus/Food Industry. Institute of Food and Agricultural Science Cooperative Extension Service. University of Florida, Gainesville.
ROLF, M. J., and LIM, H. C.   1982.   Computer control of fermentation processes. Enzyme Microbial. Technology 4: 370–380.
SIMPSON, G. S., and WILLIAMS, M. C.   1974.   Analysis of high temperature–short time sterilization during laminar flow. Journal of Food Science 39: 1047.
SMITH, C. L.   1970.   Digital control of industrial processes. Computing Surveys 2(3): 211–241.
STANDIFORD, F. C.   1963.   Evaporator performance and operation. Chemical Engineer 70(25): 164–170.
STUMBO, C. R.   1965.   Thermobacteriology in Food Processing. Academic Press, New York.
TEIXEIRA, A. A., DIXON, J. R., ZAHRADNIK, J. W., and ZINSMEISTER, G. E.   1969.   Computer optimization of nutrient retention in the thermal processing of conduction-heated foods. Food Technology 23(6): 137.
TEIXEIRA, A. A. and MANSON, J. E.   1982.   Computer control of batch retort operations with on-line correction of process deviations. Food Technology 36(4): 85–90.
TEIXEIRA, A. A., and MANSON, J. E.   1983.   Thermal process control for aseptic processing systems. Food Technology 37(4): 128–133.
THOMPSON, J. F., KRANZLER, G. A., and STONE, M. L.   1982a.   Microcomputer control for hop drying. ASAE Paper No. 82-5520. American Society of Agricultural Engineers, St. Joseph, MI.
THOMPSON, J. F., STONE, M. L. and KRANZLER, G. A.   1982b.   Modified air flow and temperature hop drying. ASAE Paper No. 82-3571. American Society of Agricultural Engineers, St. Joseph, MI.
WAGNER, J. N.   1982.   Aseptic packaging fever, aseptic drums, and aseptic bulk shipping. Food Engineering 54(1): 59.
WEIGAND, W. A., LIM, H. C., CREAGAN, C. C., and MOHLER, R. D.   1979.   Optimization of a repeated fed-batch reactor for maximum cell productivity. In Proceedings of the 2nd International Conference on Computer Applications in Fermentation Technology. Biotechnology and Bioengineering Symposium No. 9. John Wiley & Sons, New York, pp. 335–348.

5

Process Modeling and Simulation

A process model is a mathematical representation of a process in the form of an expression that correctly relates how the various product and process variables can affect the process outcome. Many food processes involve complex biochemical reactions, which are functions of time, temperature, and concentration under conditions of heat and mass transfer that are governed by various process and product conditions. The development of such models requires the expertise of food engineers who have extensive training in engineering science and mathematics in addition to basic food science.

Mathematical models for most food processes involve complex differential equations that are difficult to solve without the aid of high-speed digital computers. For this reason food process modeling and simulation remained largely of academic interest until the advent of low-cost computers and microprocessors. The capability of solving complex differential equations on small inexpensive computers allows food scientists to replace costly, time-consuming pilot-plant experiments with rapid mathematical simulations of the process at the computer terminal and allow food companies to realize quantum-leap improvements in research and process development.

THERMAL PROCESSING: A CASE STUDY

Thermal processing consists of heating food containers in pressurized steam retorts at a constant retort temperature for prescribed lengths of time. These process times are calculated as the minimum times necessary to achieve sufficient bacterial inactivation in each container to comply with public health standards and to ensure that the probability of spoilage will be less than some maximum allowable value. Associated with each thermal process is always some undesirable degradation of heat-sensitive vitamins and other quality factors. Because of these quality and safety factors, great care is taken in the calculation of these process times and in the control of time and temperature during processing to avoid either underprocessing or overprocessing. In a conduction-heated product the temperature is a distributed parameter. Because heat must penetrate from the can wall to the center, the temperature is nonuniformly distributed; that is, at any point in time the temperature is a function of location within the can. Since the thermal inactivation of bacterial spores is temperature-dependent, inactivation is occurring at different rates at different locations within the can at any point in time, while all temperatures and rates change over time at any given location.

Model Development

Although from the above description thermal processing may appear at first as a hopeless maze of complex interactions, there is a definite mathematical order to the relationships involved, which lends itself to computer modeling. An understanding of two distinct bodies of knowledge is required to appreciate the basic principles of thermal process calculations: the first is the thermal inactivation kinetics (heat resistance) of food spoilage-causing organisms, often referred to as *thermobacteriology;* and the second consists of the heat transfer considerations that govern the temperature profiles achieved within the food container during the process, commonly referred to as *heat penetration* in the canning industry.

Figure 5.1 conceptually illustrates the interdependence between the thermal inactivation kinetics of bacterial spores and the heat transfer effects in the food product. The thermal inactivation of bacteria follows first-order kinetics and can be described by a logarithmic reduction in the con-

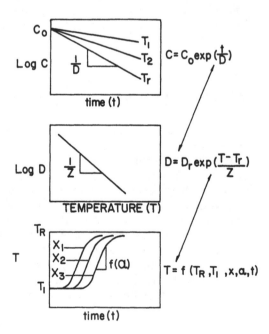

FIG. 5.1. Time- and temperature-dependence of the thermal inactivation kinetics of bacterial spores in the thermal processing of conduction-heated canned foods, where C is concentration of viable spores, D and Z are factors describing reaction kinetics, T_I and T_R are initial and retort temperature, x is any location within the container, and α is the thermal diffusivity of the product. (From Teixeira, 1978.)

centration of bacterial spores with time for any given lethal temperature, as shown in the upper family of curves in figure 5.1. These are known as *survivor curves*. The rate factor D is called the *decimal reduction time* and is expressed as the time in minutes to achieve one logarithmic cycle of reduction in concentration C. As suggested by the family of curves and specifically shown in the second curve, D is temperature-dependent and varies logarithmically with temperature. This is known as a *thermal death time curve* and is essentially linear over the range of temperatures employed in most thermal food sterilization processes. The rate constant Z used to describe this relationship is expressed as the temperature difference required for the curve to traverse one logarithmic cycle. The temperature in the food product, in turn, is a function of the retort temperature T_R, initial product temperature T_I, location within the container x, thermal diffusivity of the product α, and time t in the case of a conduction-heated food. In practice, the thermal diffusivity α is calculated from the heat penetration factor f_h, which is obtained from heat penetration curves for the product (fig. 5.2).

Thus, we can see how the concentration of viable bacterial spores during thermal processing decreases as a function of the inactivation kinetics, which are a function of temperature. The temperature, in turn, is a function of heat transfer considerations involving time, space, thermal properties of the product, and initial and boundary conditions of the process. This interrelationship is illustrated by the functional expressions given in figure 5.1.

Since the thermal properties and initial and boundary conditions remain fixed for a given process, only the temperature, as a function of time and location, varies during the process. This temperature distribution can be defined at any point in time by solving the two-dimensional partial differential equation for heat conduction in a finite cylinder (eq. 5.1). Nu-

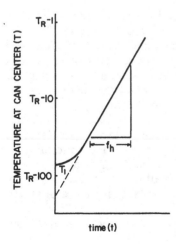

FIG. 5.2. Typical heat penetration curve describing the temperature-time profile measured at the can center, plotted on an inverted semilogarithmic scale for determining the heat penetration factor (f_h) from which apparent thermal diffusivity can be calculated. (From Teixeira, 1978.)

merical solution based on a finite difference method by high-speed digital computer is most useful for this purpose.

$$\frac{\partial T}{\partial t} = \alpha \left[\frac{\partial^2 T}{\partial r^2} + \frac{1}{r} \frac{\partial T}{\partial r} + \frac{\partial^2 T}{\partial h^2} \right] \tag{5.1}$$

where T = temperature
         t = time
         α = thermal diffusivity
         r = radial position in cylinder
         h = vertical position in cylinder

The coefficient α is defined by

$$\alpha = \frac{0.398}{\left[\dfrac{1}{R^2} + \dfrac{0.427}{H^2} \right] f_h}$$

where R = can radius
         H = one-half can height
         f_h = slope of heat penetration curve

The above discussion suggests how an overall computer model can be developed by mathematical coupling of the first-order reaction kinetics describing the thermal inactivation of bacterial spores with a numerical solution of the two-dimensional partial differential equation describing heat conduction for a finite cylinder.

The development of such a model was first described by Teixeira et al. (1969a, 1969b) and is summarized in the following section.

Numerical Solution by Finite Differences

As stated previously, the temperature within the container is neither uniform nor steady but is a function of both time and position. The concentration of viable cells is both temperature- and time-dependent and therefore can be calculated only for some point in the container at an instant in time. If such a point, however, is taken to be the center of a very small volume element relative to the entire container, then the temperature and concentration at that point can be considered representative of these values throughout the volume element surrounding it. In accord with this concept, the cylindrical container is imagined to be divided into volume elements that would appear as layers of concentric rings having rectangular cross sections. This is illustrated in figure 5.3 for the upper half of the container. Uniform surface conditions on the container allow for radial symmetry and longitudinal symmetry, which are required for this choice of subdivision.

The rate equation for bacterial inactivation can then be applied to each

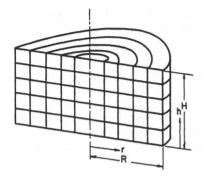

FIG. 5.3. Subdivision of a cylindrical container for numerical solution of heat conduction equation by finite differences. (From Teixeira, 1969a.)

volume element over a short time interval. Computer integration over the container volume and over process time would yield the final lethality. The rate equation for bacterial death is given by

$$- \frac{dC}{dt} = \frac{1}{D} C \tag{5.2}$$

where C is the concentration of viable bacteria at any time, D is the bacterial death rate, and t is time. Rearranging terms and integrating over a small time interval Δt, this equation becomes

$$C^{t+\Delta t} = C^t e^{-\Delta t/D} \tag{5.3}$$

where $C^{t+\Delta t}$ and C^t represent the concentration at times $t + \Delta t$ and t, respectively.

Here, D is unrestricted, and this equation holds for any rate process. However, for the purposes of this example the bacterial destruction rate is given by

$$- \frac{dD}{dT} = \frac{1}{Z} D \tag{5.4}$$

where T is the temperature and Z is the inverse slope of the thermal destruction curve. Rearranging terms and integrating over temperature, the following expression for D results:

$$D = D_r \exp \left(\frac{T_0 - T}{Z} \right) \tag{5.5}$$

where D_r is the death rate at T_0, which is usually taken to be 250°F. Thus, if the temperature is known at a given point over a short time interval, then the death rate at that point, and subsequently the concentration at that point, can be calculated.

With reference to the flow diagram in figure 5.4, a typical lethality calculation would proceed as follows. The temperature at the center of each volume element is supplied and used to calculate the death rate over a given time interval for each element by equation (5.5). Then the concentration at the end of that time interval is calculated by equation (5.3) for each element. If the process has not yet ended, this new concentration becomes the initial concentration for the next time interval, and the procedure is repeated. When the process is completed, the resulting concentrations are multiplied by the volumes of their respective elements to give the number of survivors in each element. Thus the number of survivors in the entire container will be just the sum of those in each element.

The temperature needed at each volume element after every time interval is calculated by the expression in equation (5.6):

$$
\begin{aligned}
T_{ij}^{t+\Delta t} = T_{ij}^{t} &+ \frac{\alpha \Delta t}{\Delta r^2} \left[T_{(i-1,j)} - 2T_{(ij)} + T_{(i+1,j)} \right]^t \\
&+ \frac{\alpha \Delta t}{2r\Delta r} \left[T_{(i-1,j)} - T_{(i+1,j)} \right]^t \\
&+ \frac{\alpha \Delta t}{\Delta h^2} \left[T_{(i,j-1)} - 2T_{(i,j)} + T_{(i,j+1)} \right]^t
\end{aligned}
\qquad (5.6)
$$

where Δt, Δr, and Δh are discrete increments of time, radius, and height, and i and j denote sequence of radial and vertical increments away from can wall and midplane.

This expression, which lies at the heart of the numerical computer model, is derived from the classic partial differential equation for two-dimensional unsteady-state heat conduction in a finite cylinder, written in the form of finite differences for numerical solution by digital computer.

As a framework for computer iteration, the cylindrical container is imagined to be subdivided into volume elements that appear as layers of concentric rings having rectangular cross sections, as illustrated previously in figure 5.3 for the upper half of the container. Temperature nodes are assigned at the corners of each volume element on a vertical plane, as shown in figure 5.5, where I and J are used to denote the sequence of radial and vertical volume elements, respectively.

By assigning appropriate boundary and initial conditions to all the temperature nodes (interior nodes set at initial product temperature and surface nodes set at retort temperature), the new temperature reached at each node can be calculated after a short time interval Δt that would be consistent with the thermal diffusivity of the product, which is obtained from heat penetration data. This new temperature distribution is then taken to replace the initial one, and the procedure is repeated to calculate the temperature distribution after another time interval. In this way the temperature at any point in the container at any instant in time is obtained. At the end of the process time when the steam is shut off and cooling water is admitted to the retort, the cooling process is simulated by simply chang-

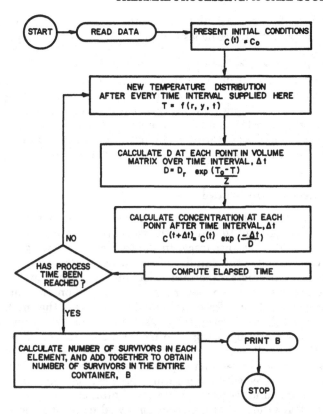

FIG. 5.4   Flow diagram describing computer program logic for calculating bacterial lethality over temperature distribution history in conduction-heated canned foods.

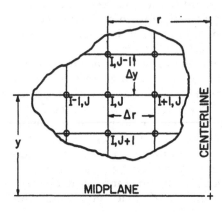

FIG. 5.5.   Labeling of grid nodes in matrix of volume elements on a vertical plane for calculating temperature distribution in canned foods by method of finite differences. (From Teixeira, 1969a.)

ing the boundary conditions from retort temperature T_R to cooling temperature T_C at the surface nodes and continuing with the computer iteration described above.

When the numerical computer model is used to calculate the process time required at a given retort temperature to achieve a specified level of sterilization, the computer follows a programmed search routine of assumed process times, which quickly converges on the precise time at which cooling should begin in order to achieve the specified lethality. In this way the model can be used to determine the process time required for any given set of constant or variable retort temperature conditions.

Validation of the Model

Once a mathematical model has been developed for a process operation, it is always important to validate the model by showing that it works with reasonable and useful accuracy. In the thermal processing of conduction-heated canned foods the validity of the model rests primarily on how accurately it can predict temperature distributions throughout the container over the process time. Whenever a model is based on an approximate numerical solution to a differential equation, as with the method of finite differences used in this case study example, the model should first be validated theoretically by using it to model a case for which an analytical (i.e., mathematically exact) solution of the differential equation exists. The model is theoretically valid if temperatures predicted by the finite difference approximation agree with temperatures predicted by the analytical solution at a given time and location within the container.

The partial differential equation for two-dimensional heat conduction in a finite cylinder, equation (5.1), has an exact analytical solution for the case in which the boundary temperature (retort temperature) is given an initial step increase at time zero and held constant over process time. An analytical solution also exists for the case in which the boundary temperature varies as a sinusoidal function of time. The use of both these solutions for predicting temperatures in conduction-heated foods has been reported by Hayakawa (1964). Since the boundary temperature in the finite difference model can be specified as any desired function of time, it can be easily defined to simulate either of these two conditions.

A small computer program can be written to execute each of these analytical solutions for predicting the temperature at a given time and location. The temperatures obtained can then be compared with those predicted by the finite difference model at the same time and location for the same boundary and initial conditions, can size, and thermal properties specified for the analytical solutions. Tables 5.1 and 5.2, respectively, show the results of such an exercise for the case in which the retort temperature was held constant at 250°F and for the case in which it varied as a sinusoidal function of time between a maximum of 265°F and a minimum of 226°F. The initial product temperature was assumed uniform at 160°F in all cases.

TABLE 5.1. Comparison of Temperature Distribution Histories from a Computer Model with Hayakawa's Analytical Solutions for Constant Surface Temperature (250°F)

Heating Time (min)	Method	Radial Location on Midplane (r/R)					
		0.0	0.2	0.4	0.6	0.8	1.0
1.0	Analytical	160.0	160.0	160.0	160.0	165.2	250.0
	Computer	160.0	160.0	160.0	160.2	167.1	250.0
10.0	Analytical	161.5	163.3	169.4	185.6	214.5	250.0
	Computer	161.9	163.5	169.8	185.6	214.1	250.0
30.0	Analytical	195.5	198.6	206.9	219.7	234.9	250.0
	Computer	195.6	198.5	206.8	219.6	234.8	250.0
84.0	Analytical	242.7	243.1	244.3	246.0	248.0	250.0
	Computer	242.6	243.0	244.2	246.0	248.0	250.0

(*Source:* Teixeira, 1971.)

Results are reported for six different radial locations on the midplane of the container at four different points in time throughout the process. Since agreement between analytical and computer model predictions is within a fraction of a degree at nearly all points, the model can be assumed to be theoretically valid.

The next step in process model validation is to determine how closely computer-predicted temperatures can agree with temperatures measured experimentally by thermocouples mounted in a can of product undergoing an actual conduction heating process in a pilot-plant retort. Figure 5.6 shows schematically how a large No. 10 can was instrumented for such an experiment in order to have thermocouple junctions placed at different radial locations on its midplane. The can was filled with a 5 percent bentonite suspension as a conduction heating model system. The thermal diffusivity was calculated from the heat penetration factor f_h, which was obtained from repeated heat penetration tests, as shown earlier in figure 5.2 and equation (5.1). Experiments were carried out for two different sets of boundary conditions: in one experiment the retort temperature was held constant at 250°F and in the other it was increased stepwise as a function

TABLE 5.2. Comparison of Temperature Distribution Histories from a Computer Model with Hayakawa's Analytical Solutions for a Sinusoidal Surface Temperature Policy (226–265°F)

Heating Time (min)	Method	Radial Location on Midplane (r/R)					
		0.0	0.2	0.4	0.6	0.8	1.0
1.0	Analytical	160.0	160.1	160.1	160.1	165.0	245.9
	Computer	160.0	160.0	160.0	160.2	166.7	245.9
10.0	Analytical	160.4	162.8	169.1	186.1	215.5	254.9
	Computer	161.8	163.4	169.6	185.4	214.9	254.9
30.0	Analytical	195.2	200.3	210.2	225.6	244.7	265.0
	Computer	196.7	200.3	210.1	225.4	244.7	265.0
84.0	Analytical	246.1	244.0	242.3	239.0	233.8	226.0
	Computer	244.5	244.0	242.5	239.3	233.9	226.0

(*Source:* Teixeira, 1971.)

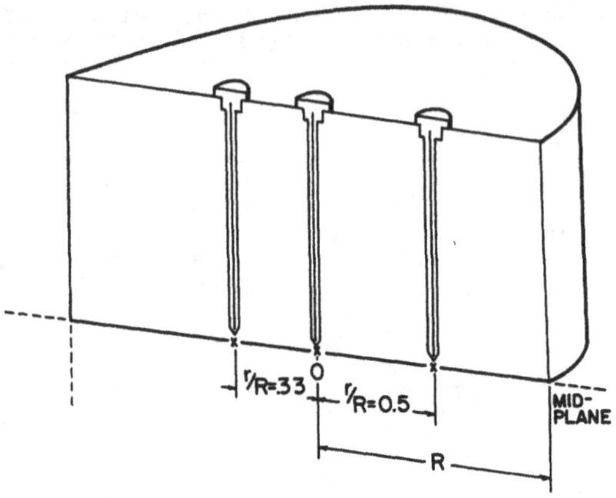

FIG. 5.6. Schematic diagram of a No. 10 can showing location of thermocouples for heat penetration tests to compare computer-predicted temperatures with measured temperatures at the same location. (From Teixeira, 1971.)

of time to show the application of the model for time-varying boundary conditions.

The measured and predicted temperatures from these experiments are shown in Tables 5.3 and 5.4 for constant and time-varying boundary conditions, respectively. Results in both cases show temperature agreement at all points to be well within the limits of expected thermocouple accuracy and experimental error. In most cases, temperature differences are less than 1°F. These experimental results, combined with the results from analytical solutions discussed earlier, provide strong support for the validity of the model in predicting accurately the temperature distribution in conduction-heated canned foods at any specified boundary process condition.

At this point the model has only been shown to be valid for predicting temperatures; its ability to accurately predict the results of biochemical reactions with temperature-dependent reaction kinetics remains to be demonstrated. In his textbook on thermobacteriology in food processing, C. R. Stumbo (1965) described an empirical method for calculating the integrated lethality in a conduction-heated canned food. The method is

TABLE 5.3. Comparison of Temperature Distribution Histories between Computer Predictions and Experimental Measurements for a Constant Retort Temperature

Heating Time (min)	Method	Radial Location on Midplane (r/R)			
		0.00	0.33	0.50	1.0
30	Computer	83.9	93.1	111.0	250.0
	Experiment	83.0	92.5	110.5	250.0
90	Computer	153.4	167.0	184.0	250.0
	Experiment	152.0	166.5	184.0	250.0
150	Computer	206.0	212.5	220.4	250.0
	Experiment	206.0	212.5	220.0	250.0
210	Computer	230.5	233.4	236.9	250.0
	Experiment	231.0	233.5	237.0	250.0

(*Source:* Teixeira, 1971.)

TABLE 5.4. Comparison of Temperature Distribution Histories between Computer Predictions and Experimental Measurements for a Step-Increasing Retort Temperature Policy

Heating Time (min)	Method	Radial Location on Midplane (r/R)			
		0.00	0.33	0.50	1.00
30	Computer	90.5	100.2	117.1	230.0
	Experiment	90.5	101.0	115.0	230.0
90	Computer	152.3	164.1	178.9	240.0
	Experiment	151.5	164.5	179.0	240.0
150	Computer	199.6	206.2	214.6	250.0
	Experiment	199.5	206.5	214.5	250.0

(*Source:* Teixeira, 1971.)

based on a description of isothermal regions in the can (regions experiencing the same temperature at any point in time). By defining the fraction of can volume enclosed by these regions, the lethality achieved in each region could be calculated for the temperature-time history of that region by the classical Ball formula (Ball, 1938) and integrated over all regions to obtain an approximation of the total lethality achieved throughout the container for a given thermal process. Stumbo included a number of sample problems in his text to provide exercise for mastering this method of calculation.

By using the process conditions given in Stumbo's sample problems to specify initial and boundary conditions for the computer model, Teixeira et al. (1975a) showed how the computer model could be used to calculate the total integrated process lethality for the same size can and heat penetration data given in each of Stumbo's sample problems and the results could be compared for agreement between both methods. Table 5.5 shows the results obtained by both methods for five sample problems taken from Stumbo (1965). The results are the numbers of survivors remaining at the end of the thermal process for each sample problem. The first column lists

TABLE 5.5. Solutions to Sample Problems

Problem Number	Stumbo's Solution	Computer Solution of Manson and Zahradnik (1967)	Solution by New Computer Technique Using a (10 × 10) Matrix	Solution by New Computer Technique Using a (5 × 5) Matrix
1	0.001	0.001037 (0.001)*	0.0006630 (0.001)*	0.002346
2	0.01	0.01486 (0.01)*	0.01300 (0.01)*	0.02458
3	0.01	0.01226 (0.01)*	0.01128 (0.01)*	0.02355
4	0.01	0.01060 (0.01)*	0.008320 (0.01)*	0.01080
5	3.3	3.309 (3.3)*	2.090 (2.1)*	2.348

*Values rounded to the number of significant figures quoted by Stumbo (1965).
(*Source:* Teixeira et al., 1969a.)

Stumbo's solutions; the second column lists the solutions obtained by Manson and Zahradnik (1967), who programmed Stumbo's method for a high-speed digital computer; and the third and fourth columns list solutions obtained from the computer model developed in this case study. Note that, with the exception of the last problem, the results agree to the first significant figure, which is consistent with the significance of Stumbo's values.

The difference in results beyond the first significant figure can be attributed to the fact that Stumbo's temperature distributions were obtained by describing a number of iso-j regions (i.e., regions having the same heating lag) extending only part way through the container. The computer model developed in this case study used 200 volume elements in the entire container, and the temperature at the center of each element was calculated every eighth of a minute. It was also noted that when larger volume elements were taken, the results tended toward those of Stumbo. This may be verified by inspecting the fourth column of Table 5.5, where the results are listed for a volume element matrix of only five radial and five vertical increments. The significant data describing each of the five sample problems is given in Table 5.6.

The chief significance of this comparison is to show that the results are reasonably consistent with those obtained by Stumbo's method and that the computer model may be used with a high degree of confidence. Moreover, the versatility and possible applications of the model cannot be overemphasized. Specifically, the integrated effect on any variable of a thermal process that follows first-order kinetics with the rate varying logarithmically with temperature can be calculated. Like bacterial lethality, thiamine degradation in canned foods during thermal processing involves such a reaction. Thus, by inputting the appropriate rate data,

TABLE 5.6. Table of Significant Data for Sample Problems

| Parameter | Input Data for Sample Problems | | | | |
	1	2	3	4	5
A	9000	2850	2850	1275	1275
TR	250	250	250	240	235
TI	160	160	160	160	160
U	255	83.5	96	131	131
fh	200	60	60	55	55
Dr	3.2	3.0	3.0	3.0	3.0
Z	14	18	12	16	16
Can size*	603 × 700	307 × 409	307 × 409	303 × 411	303 × 411

*These numbers represent the diameter and height of the container, respectively (e.g., the container used in problem 2 has a diameter of 3⁷⁄₁₆ in and a height of 4⁹⁄₁₆ in). The other parameters are defined as follows:

A = initial spore load in container
TR = retort temperature
TI = initial food temperature
U = process time
fh = the inverse slope of the centerpoint heating curve
Dr = the death rate at 250°F
Z = the inverse slope of the thermal-death-time curve
(*Source:* Teixeira et al., 1969a.)

thiamine retention could be calculated with this new computer model for various thermal processes.

In fact, the thermal degradation kinetics for thiamine in different foods had been studied and reported by Felliciotti and Esselen (1957). Using the D and Z values reported in that work for the thermal degradation of thiamine in pea puree, Teixeira et al. (1975b) carried out the final step in validating the computer model developed in this case study example by comparing computer model predictions of thiamine retention with experimental results from laboratory assays of thiamine content in cans of pea puree subjected to different thermal processes.

Four different retort processes were specified by choosing different retort temperatures and corresponding process times to make them all equivalent with respect to their sterilization capacity for spores of *Bacillus stearothermophilus*, whose heat resistance is characterized by $D_{250} = 4$ min and $Z = 18°F$. Constant retort temperatures of 240, 250, and 260°F were specified for three of the processes. The required retort heating and cooling times to achieve a five-logarithmic-cycle reduction in any initial spore population were calculated by Stumbo's (1965) method and checked with the computer model. The fourth process was specified with a stepwise increase in retort temperature from 230 to 240, to 250 to 260°F, with 20 minutes between each step change. The process time was determined by the computer model on the basis of the same five-logarithmic-cycle reduction in spore population required of all four processes. Times and temperatures for each of these processes are summarized in Table 5.7.

Since Felliciotti and Esselen (1957) did not establish thiamine destruction curves for 250°F, it was first necessary to convert the D value from one of the temperatures they used (a D value of 202.3 min at 246°F) to a D value of 165.6 at 250°F. The thiamine retention for these processes was computed by using destruction rates defined by a D_{250} value of 165.6 and a Z value of 46.

The thiochrome method of thiamine assay described by the Association of Vitamin Chemists (1951) was used in this study. In order to experi-

TABLE 5.7. Description of Retort Processes Prescribed for Thiamine Assay Comparisons*

Process Number	Retort Temperature (°F)	Process Time (min)
1	250	85.0
2	240	136.0
3	260	65.0
4		87.0

*All processes have equal sterilizing value with respect to spores of *B. stearothermophilus*.
(*Source:* Teixeira et al., 1975b.)

mentally determine the thiamine retention associated with any one process, the assay procedure was conducted simultaneously on six individual samples of pea puree, three of which were taken from cans that had been retort-processed in the pilot plant while the other three (control samples) were taken directly from the freshly prepared batch immediately prior to canning. Sufficient assay solution was prepared from each sample to permit four replicate readings on the Coleman photofluorometer plus one blank determination for each sample. This resulted in a total of 12 replicate readings to represent the thiamine concentration in the three processed samples and 12 to represent the thiamine concentration in the three control samples.

When the assays were completed, the replicate readings were treated statistically to determine standard deviations and confidence limits on the mean values. The percent thiamine retention was then calculated directly from the mean values of the scale readings for the processed samples and control samples, since the meter deflections on the photofluorometer were linearly proportional to the thiamine concentrations for the concentration range used.

In addition to comparing computer predictions with laboratory analyses, thiamine retentions were also calculated in Jen et al.'s (1971) modification of Stumbo's method. This method, however, applied only to the three thermal processes in which the retort temperature was held constant.

The results from all three methods are compared in Table 5.8. Agreement between calculated values and laboratory assay results is, in all cases, well within expected error and confirms the validity of both the computer model developed in this case study example and of Stumbo's method as modified by Jen et al. (1971). A similar confirmatory comparison was reported by Jen et al. (1971), which included the assay determinations reported by Hayakawa (1969) and predictions of percent thiamine retention by the methods of Ball and Olson (1957), Hayakawa (1969), Teixeira et al. (1969a), and Jen et al. (1971).

Applications and Limitations of the Process Model

The material presented thus far in this chapter has shown how a mathematical process model was developed for computer simulation of the thermal processing of conduction-heated canned foods; it also has shown how the model was proved valid for predicting the temperature distribution history in the can for processes with both constant and time-varying retort temperatures, as well as for predicting the outcome of biochemical reactions with temperature-dependent kinetics, such as the inactivation of bacterial spores and the degradation of thiamine. Such a model can become a very powerful tool in the hands of the food scientist or engineer, since it can predict the effect of various conditions imposed on the process, product, or containers on the final process outcome.

In research applications the model can be used to establish new processes, to evaluate existing processes, and to determine optimum process conditions to achieve minimum quality degradation while meeting the required degree

TABLE 5.8. Comparison of Thiamine Retention in 303 × 406 Cans of Pea Puree as Calculated by Modified Stumbo Method, Teixeira's Computer Model, and Laboratory Assay Determinations for Various Thermal Processes

		Percent Thiamine Retention by		
Process Number	Process Description	Modified Stumbo Method	Computer Model	Laboratory Assay
1	85 min at 250°F	48.7	49.2	50.7
2	136 min at 240°F	41.0	41.6	42.3
3	65 min at 260°F	47.3	48.8	51.3
4	Step increase regime		50.2	53.5

(*Source:* Teixeira et al., 1975b.)

of bacterial spore inactivation. In process control applications the model can be used as part of a computer-based control system to make on-line corrections in response to unexpected temperature deviations and thereby ensure that products have achieved the specified level of sterilization when the process is completed. In quality assurance applications the model can be used to quickly evaluate a batch of product that experienced any deviation in process conditions from the established process so that product safety and quality may be ensured before release for shipment. Specific applications of this model in process optimization will be described as a continuing case study example in Chapter 6.

The case study example in this chapter has focused on the development of one mathematical model for a specific unit operation, thermal processing of canned food, which makes use of a numerical solution by finite differences of the differential equation for heat transfer. Other mathematical models developed to simulate the thermal processing of canned foods make use of different methods for solving the differential equation. Hayakawa (1969), for example, makes extensive use of the exact analytical solution of the differential equation in combination with carefully selected dimensionless parameters.

Although the method of finite differences provides added flexibility to accommodate irregularly time-varying boundary conditions, it shares the limitations of analytical solutions in that it is practical only for regular shapes such as slabs, cylinders, and spheres. The finite element method is an alternative to the finite difference method, which can accommodate almost limitless irregularities, such as nonuniform initial temperature distribution, nonlinear and nonisotropic thermal properties, and irregularly shaped bodies, as well as time-dependent boundary conditions. The finite element method makes use of matrix algebra for solving the complex differential equation for heat flow and is totally different from the method of finite differences, although both methods involve subdivision of the body into finite sections. A finite element method for use in modeling the thermal processing of foods is given in Naveh et al. (1983). The price usually paid for the extra capability of the finite element method is the added effort to set up the model and input data, which is followed by much longer computer execution time.

FREEZING

Food preservation by freezing is at least equal, if not greater, in economic importance to the U.S. food industry as food preservation by canning, and the use of mathematical models for computer simulation of the freezing process is of great value in freezing system design. The ability to accurately predict freezing times for various product and process conditions is a fundamental requirement for the design of commercial food freezing systems. The use of reliable computer simulation models can go a long way in minimizing the need for costly laboratory and pilot-plant experiments for this purpose.

Predicted freezing times for foods can be approximated by traditional analytical methods that do not require numerical solution by computer simulation. The most popular of the analytical methods is Plank's equation:

$$t_F = \frac{\rho L}{T_F - T_x} \left[\frac{Pa}{h} + \frac{Ra^2}{k} \right] \tag{5.7}$$

where
- t_f = freezing time
- ρ = product density
- L = latent heat of fusion
- T_F = initial freezing temperature
- T_x = temperature of freezing medium
- P and R = constants depending on geometry
- a = characteristic dimension
- h = surface heat transfer film coefficient
- k = thermal conductivity of frozen product

The chief limitations to Plank's equation are outlined by Heldman (1983):

The latent heat of fusion L must be estimated as an approximate value based on the initial moisture content and heat of fusion for pure water.
The equation does not account for sensible heat removal above the initial freezing point T_F, nor for heat removal from the frozen product.
The selection of a thermal conductivity k value for the frozen product is difficult since this property is temperature-dependent.

Various modifications of Plank's equation have been proposed. The first was by Nagaoka et al. (1955) to account for sensible heat removal, and the most recent by Cleland and Earle (1982) to account for a variety of product shapes. In general, these modifications deal primarily with improvements in estimates and can be accurate only under limited conditions.

In order to account for all the unique features and irregularities that are inherent in most realistic food freezing applications, the appropriate mathematical expressions for heat transfer with temperature dependent thermal properties must be solved numerically by using computer simulation. According to Heldman and Singh (1981) for the case of one-dimensional heat conduction, the following partial differential equation:

$$\frac{\partial T}{\partial t} = \frac{\partial}{\partial x}\left[\alpha\frac{\partial T}{\partial x}\right] \qquad (5.8)$$

would be solved numerically by using appropriate initial conditions and boundary conditions to account for heat transfer at the product surface.

Equation (5.8) is similar in form to any one of the terms shown in equation (5.1) for the two-dimensional partial differential equation for heat conduction in a finite cylinder. One important difference is that the thermal diffusivity α is not shown outside the partial derivative expression as a constant but is treated as a variable. Equation (5.8) can be solved numerically as long as a function describing the thermal diffusivity as a function of temperature is known. This function would then become part of the expression used in writing equation (5.8) in finite difference form. Alternatively, it would become a matrix multiplier in the method of finite elements as described by Purwadaria and Heldman (1982).

The thermal diffusivity can be defined in the following manner (Heldman and Singh, 1981):

$$\alpha\,(T) = \frac{k(T)}{\rho(T)\,C_P(T)} \qquad (5.9)$$

where thermal conductivity k, density ρ, and specific heat C_P are functions of temperature during the freezing process. Discussions of the temperature dependence of each of these properties, along with specific temperature-dependent expressions for their use in numerical solutions, are given in Heldman and Singh (1981).

Numerical solutions by computer simulation of the heat conduction problems in food freezing processes have been explored by Bonacini and Comini (1973), Bonacini et al. (1973, 1974), Hohner and Heldman (1970), Heldman (1974), and Heldman and Gorby (1975a, 1975b).

In addition to improved accuracy in predicting freezing times, computer simulation models also make it possible to quickly study the effect of various product and process factors on freezing rates without costly laboratory or pilot-plant experiments. Table 5.9, taken from Heldman (1983), shows a comparison of freezing times for strawberries with various diameters as predicted by four different methods. The first three are analytical methods reflecting Plank's equation and modifications to it. The fourth is a numerical method by computer simulation developed by Heldman (1982). The results clearly show how use of the Plank and modified Plank equations can lead to serious underestimation of freezing time, particularly as the food particle size increases. Table 5.10 has been included to show the input data that were used in the analytical and numerical freezing time prediction methods shown in Table 5.9.

Figures 5.7 through 5.11 were taken from Heldman (1983) to illustrate the power of computer simulation models in studying factors affecting freezing rates. Figure 5.7 shows the predicted temperature history gen-

TABLE 5.9. Comparison of Freezing Times for Strawberries with Various Diameters Predicted by Using Four Methods of Computation

	Freezing Times (min) for Different Product Diameters		
Method	1.0 cm	2.0 cm	3.0 cm
Plank	3.60	7.77	12.49
Modified Plank	3.44	7.40	11.91
Cleland-Earle	4.74	10.41	17.03
Computer simulation	4.20	13.00	21.90

(*Source:* Heldman, 1983.)

TABLE 5.10. Values of Input Parameters for Strawberries Used in Freezing Time Predictions by Analytical and Numerical Methods

	Parameter Value for Strawberries	
Parameter	Analytical	Numerical
Density, unfrozen	1040 kg/m^3	1040 kg/m^3
Density, frozen	960 kg/m^3	—
Water content	89.3%	89.3%
Thermal conductivity, unfrozen	0.54 W/m/K	0.54 W/m/K
Thermal conductivity	2.08 W/m/K	—
Specific heat, unfrozen	3.93 kJ/kg/K	3.93 kJ/kg/K
Latent heat	374.4 or 297.6 kJ/kg	—
Freezing medium temperature	−35°C	−35°C
Surface heat transfer coefficient	70 W/m^2/K	70 W/m^2/K
Initial temperature	10°C	10°C

(*Source:* Heldman, 1983.)

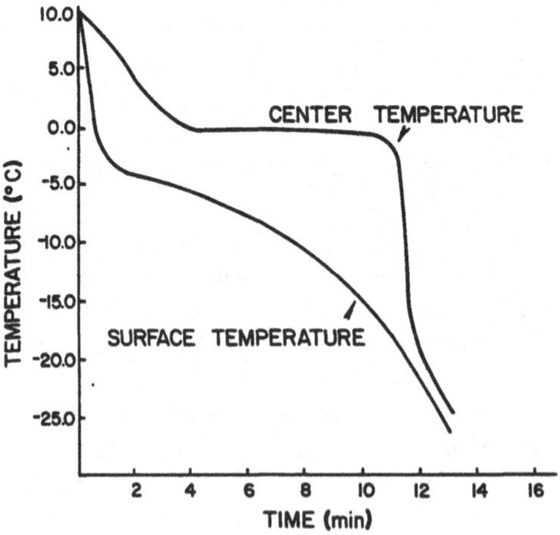

FIG. 5.7. Predicted temperature history at the surface and center of a 2-cm strawberry (T_i = 10°C, T_∞ = −35°C, h = 70 W/m^2/K, a = 2 cm). (From Heldman, 1983.)

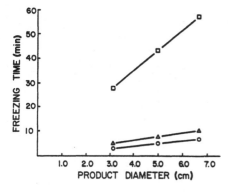

FIG. 5.8. Influence of product size on predicted freezing time (For □, h = 22.7 W/m²/K; for △, h = 170.35 W/m²/K; for ○, h = 340.7 W/m²/K). (From Heldman, 1983.)

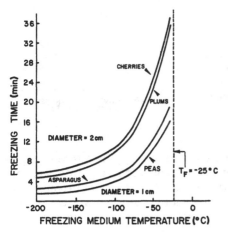

FIG. 5.9. Influence of freezing medium temperature on predicted freezing times of fruits and vegetables (T_1 = 10°C, h = 25 W/m²/K). (From Hsieh et al., 1977.)

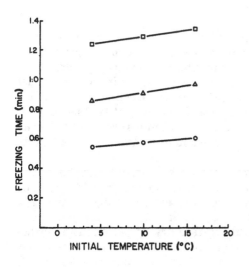

FIG. 5.10. Influence of initial product temperature on predicted freezing time (For □, diameter = 2.54 cm; for △, diameter = 1.91 cm; for ○, diameter = 1.27 cm; T = 196°C; h = 170.35 W/m²/K). (From Heldman, 1983.)

153

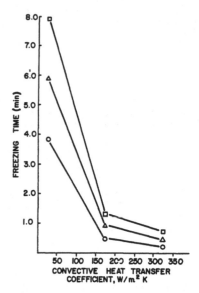

FIG. 5.11. Influence of convective heat transfer coefficients on predicted freezing times (for □, diameter = 2.54 cm; for △, diameter = 1.91 cm; for ○, diameter = 1.27 cm; $T_1 = 10°C$; $T_\infty = -196°C$). (From Heldman, 1983.)

erated by the model for a given set of conditions. Such temperature histories can be compared with actual temperatures measured in laboratory experiments to validate the accuracy of the model. The effects of product size, freezing medium temperature, initial product temperature, and convective heat transfer coefficient on predicted freezing times are shown in figures 5.8, 5.9, 5.10, and 5.11, respectively.

DRYING

Food preservation by dehydration immediately follows canning and freezing among the food processing operations of greatest economic importance to the U.S. food industry. Since the primary objective in food dehydration is the removal of water, mathematical models of the drying process must include mass transfer of moisture by diffusion from the interior to the surface of the food particle being dried, in addition to heat transfer effects. Also, the degradation of quality factors such as vitamins or pigments during processing is often a function of moisture content as well as time and temperature.

A classic example of a mathematical model for describing the drying behavior of a food product by computer simulation was that developed by Saguy et al. (1980) to predict beet pigment retention during air drying of beet slices. The model was based on application of Fick's law for one-dimensional moisture flow in the falling-rate drying period. The mathematical model and the data generated from kinetic studies of temperature- and moisture-sensitive red beet pigments (betanine and vulgaxanthin-I) were combined in a computer program to simulate and predict beet pigment retention as a function of the process variables. The concept and approach are very similar to those of the model for predicting thiamine retention and bacterial lethality as a function of thermal process variables developed earlier in this chapter.

According to Saguy et al. (1980), Fick's law for unidimensional moisture flow in the region $0 \leqslant x \leqslant L$ was applied:

$$\frac{dm}{dt} = D\frac{d^2m}{dx^2} \qquad (5.10)$$

where D = diffusion coefficient (cm²/s)

L = slab thickness (cm)

m = moisture content (g water per gram of solids)

t = time (s)

x = distance from the surface of the slab (cm)

The boundary and initial conditions were specified as

$$m = m_i \text{ at } t = 0 \text{ for all } x$$
$$m = m_e \text{ at } x = 0, L \text{ for all } t > 0$$

where m_i = initial moisture content, and m_e = equilibrium moisture content. The moisture distribution m_x and the average moisture content m for the first drying period can be expressed as

$$M_x = m_e + \frac{4}{\pi}(m_i - m_e)$$

$$\sum_{n=0}^{\infty} \frac{(-1)^n}{(2n + 1)} \cos\left[\frac{2n + 1}{L}(\pi x)\right] \exp\left[-\frac{\pi^2(2n + 1)^2}{L^2}\right] \qquad (5.11)$$

$$E \equiv \frac{m - m_e}{m_i - m_e} = \frac{8}{\pi^2}\sum_{n=0}^{\infty}\frac{1}{(2n + 1)^2}\exp\left[-\frac{\pi^2(2n + 1)^2}{L^2}Dt\right] \qquad (5.12)$$

In the falling-rate drying period, where $E \leqslant 0.6$, equation (5.12) reduces to:

$$E = \frac{8}{\pi^2}\exp(-\pi^2 Dt/L^2) \qquad (5.13)$$

The diffusion coefficient D may be expressed by a general Arrhenius-type equation:

$$D = D_o\exp(-E_a/RT) \qquad (5.14)$$

where D_o = frequency factor (cm²/sec)

E_a = activation energy (cal/mole)

R = gas constant (1.987 cal/mole/K)

T = absolute temperature (K)

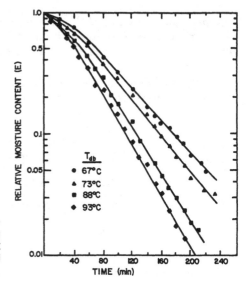

FIG. 5.12. Relative moisture content E versus time showing drying behavior of beet slices at different dry-bulb air temperatures. (From Saguy et al., 1980.)

The values of D_o and E_a are obtained from actual drying rate experiments at different dry-bulb air temperatures, as shown in figures 5.12 and 5.13.

The rate of pigment loss under dynamic dehydration conditions, in which temperature and moisture are changing continuously, was expressed by Saguy et al. (1978a) as:

$$\frac{dc}{dt} = -k\,(M,T)\,C^n \tag{5.15}$$

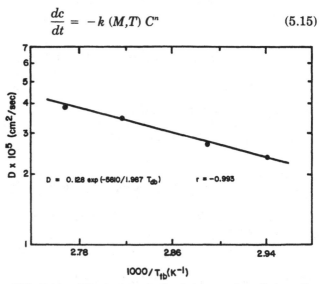

FIG. 5.13. Effect of dry-bulb air temperature (t_{db}) on the diffusion coefficient D of water in beet slices. (From Saguy et al., 1980.)

where c = pigment concentration
    k = reaction rate constant
    m = moisture content
    n = order of reaction
    t = time
    T = absolute temperature

Equation (5.13) can be integrated to give:

$$\int_{c_0}^{c} \frac{dc}{c^n} = -\int_{0}^{t} k\,(m,t)dt \tag{5.16}$$

where c_0 = initial concentration and c = concentration at time t.

For a first-order reaction ($n = 1$) the integral on the left side of equation (5.16) assumes the value in ln (c/c_0) at any given time t. Substituting

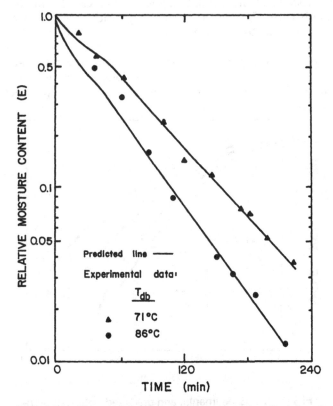

FIG. 5.14.   Comparison between theoretical (predicted) and experimental drying curves showing relative moisture content E versus time for beet slices. (From Saguy et al., 1980.)

the general Arrhenius-type equation to describe the pigment loss kinetics, equation (5.14) can be rewritten as:

$$\ln (c/c_0) = - \int_0^t k_0 (m) \exp [-E_a (m)/RT] \, dt \qquad (5.17)$$

where c_0 = initial concentration and c = concentration at time t.

Pigment retention at any time can be calculated by integrating equation (5.17) numerically through computer simulation. The functions $k_0(m)$ and $E_a(m)$ were found from the kinetic studies on beet pigment degradation carried out by Saguy et al. (1978a, 1978b). Thus, equation (5.17) permits calculation of pigment retention as a function of temperature, moisture content, and time.

Figures 5.14 to 5.16 were taken from Saguy et al. (1980) to show how the model was validated. Computer-predicted drying curves are compared with experimental data for two different temperatures in figure 5.14, while computer-predicted betanine retention and vulgaxanthin-I retention are compared with experimental data points in figures 5.15 and 5.16, respectively. In all three figures the computer-predicted results are shown as smooth curves, and the experimental results are shown as specific data points.

Often, the type of drying equipment or equipment configuration system used in the drying process will dictate various aspects of the simulation model. For example, a simulation model for a distributed-parameter tunnel

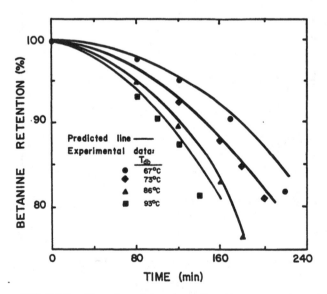

FIG. 5.15.   Experimental and predicted betanine retention during air drying of beet slices (final moisture content = 0.01 g water per gram of solids). (From Saguy et al., 1980.)

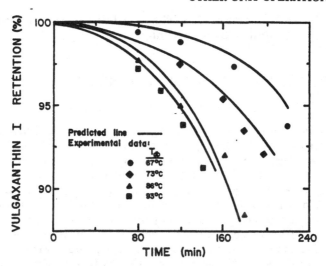

FIG. 5.16.   Experimental and predicted vulgaxanthin-
I retention during air drying of beet slices (final moisture
content = 0.01 g water per gram of solids). (From Saguy et al., 1980.)

drying system is described by Bertin et al. (1980). Troeger and Butler (1979)
describe a simulation model for solar drying of peanuts, while Barrett et
al. (1981) describe the simulation of low-temperature wheat drying, and
Bloore and Boag (1982) describe a simulation model for a tall-form spray
drier. Simulation models for fixed-bed grain dryers are reported by Sabbah
et al. (1979), Fraser and Muir (1981), and Thompson et al. (1982).

OTHER UNIT OPERATIONS

In addition to canning, freezing, and drying; process models have been
developed and reported in the literature for computer simulation of other
unit operations important to the food industry, such as evaporation, fermentation, extrusion, and cooling. Models for the simulation of multiple-
effect evaporators have been described by Radovic et al. (1979) and Chen
et al. (1980). Computer simulation of the vacuum evaporation of tomato
paste is described by Lima Hon et al. (1979), and Neelakantan and Mukesh
(1979) describe a computer model for a continuous evaporative crystallizer.

A broad coverage of computer applications in fermentation technology
is given by Armizer (1979), while computer simulation models for batch
and continuous fermentation of ethanol are described in detail by Meiering
and Subden (1983). The computer simulation of a single-screw cooking extruder is described by Pisipati and Fricke (1979). Computer simulation of
the forced convection cooling of sugar beets is described by Holdredge and

Wyse (1982), while Arce et al. (1983) describe the modeling of beef carcass cooling using a finite element technique.

COMPUTER-AIDED DESIGN IN PROCESS FLOWSHEETS

This section steps away from the modeling of individual unit operations and addresses the use of computers for simulation of a total process, made up of a sequence of unit operations on various product and by-product flow streams that comprise a complete manufacturing process. The flowsheet is the universal tool by which a manufacturing process is described. In its simplest form it is a schematic block diagram in which the unit operations are represented by boxes connected by lines representing the flow of product as it moves from one operation to another. In a more complex form flowsheets also show the mass flow rate of each product stream as it enters and exits from each unit operation box. Energy flow streams are also shown and quantified as amounts of fuel, steam, or electric power needed by each unit operation. Once this information is available on a flowsheet, engineers can determine the economics of a process. The mass and energy balance calculations that must be applied to each unit operation are tedious and time-comsuming, so that considerable time and effort is often required to determine the effect of small changes on the overall process economics.

Computer software packages are now available for application to food processing operations of a computer flowsheet calculation system for steady-state mass and energy balances (Drown and Petersen, 1983). It is advantageous to use a modular or building block approach to computer calculations of mass and energy balances because all food processing operations consist of the same basic unit operations (mixing, flow splitting, heating, cooking, drying, etc.) Food processes vary primarily in the way these basic unit operations are connected and in specific operating conditions within the unit operation. This fact suggests that if computer blocks are written to represent the material and energy balances for the basic unit operations and if the computer system can connect these blocks in alternative configuratons, then a process engineer can compute material and energy balances for different operating systems without writing new programs.

The petroleum and petrochemical industries have used a modular approach to material and energy balance computations for over 20 years, and the pulp and paper industry has been using a modular approach for the past 12 years. The most widely used modular computer system for mass and energy balance calculations in the pulp and paper industry is GEMS (an acronym for general energy and material balance system) (Edwards and Baldus, 1979). It has been under development at the University of Idaho for over 15 years. GEMS consists of a large number of blocks describing the various unit operations in pulp and paper mills, plus an executive program to connect the blocks and keep appropriate records. GEMS has been expanded to include unit operations common to the food processing industry. The first application to be described is the recovery of peel oil from oranges during juice extraction. Alternative flowsheets were analyzed to improve oil recovery and to minimize operating problems in existing equipment. The second application is concerned with optimizing the performance of a potato blanching process for controlling sugar content with minimum energy consumption.

Orange Peel Oil Recovery System Simulation

The objective of this simulation was to develop process alternatives to an existing orange peel oil recovery system that were more economically feasible and operationally reliable. The process flowsheet of the oil recovery system is shown in figure 5.17. The desludger and polisher shown are both centrifuges that separate the oil emulsion into solids, water, and oil. The aqueous stream from the desludger is recycled back to the extractor.

The oil recovery from such a system was good but there were operating problems. Typical oil recovery was 56.8 percent under existing process conditions. The recycle stream had a fairly high content of suspended solids, which clogged the holes in the spray rings of the extractor on a regular basis. A successful filtering scheme was yet to be developed to solve this problem, which resulted in an abundance of downtime and clogged spray rings. Therefore a nonrecycle system was preferred by many operators; however, the nonrecycle system produces 10 percent less oil, for only a 45.4 percent oil recovery. The recycle system was used as a base case; the calculational block diagram for its computer simulation is shown in figure 5.18.

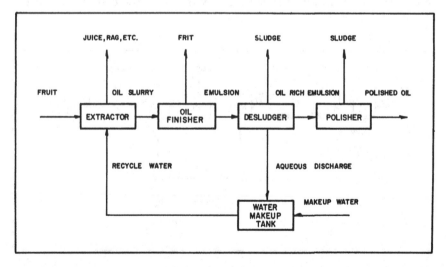

FIG. 5.17. Oil recovery system material balance base case. (From Drown and Petersen, 1983.)

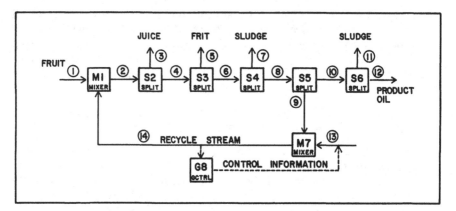

FIG. 5.18.   GEMS base case block diagram of oil recovery system. (From Drown and Petersen, 1983.)

The alternative case, based on recycling the aqueous discharge to the desludger instead of the extractors, is shown in figure 5.19. The desludging centrifuges were not being operated at nearly their design capacity, which is also their optimum operating condition. Therefore the aqueous discharge was recycled the the desludger to bring it up to capacity and permit it to operate at optimum efficiency. Fresh water could then be used in the spray rings, eliminating the plugging problem. This equipment flowsheet arrangement is shown in figure 5.20 and the result is a 4 percent loss in oil recovered compared with the base case recycle system, but the savings in downtime and maintenance could easily make up the cost of the lost oil.

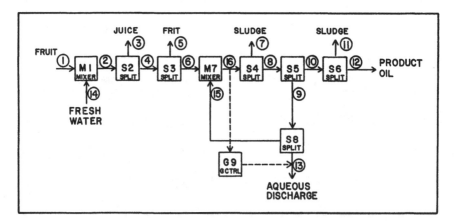

FIG. 5.19.   GEMS alternative case block diagram of oil recovery system. (From Drown and Petersen, 1983.)

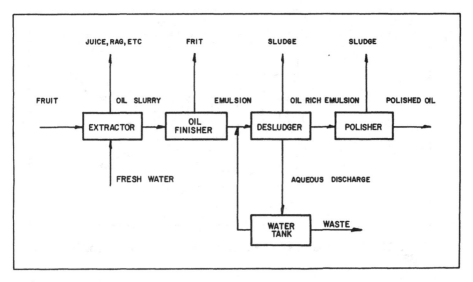

FIG. 5.20.   Oil recovery system alternate equipment flowsheet. (From Drown and Petersen, 1983.)

A reduction of 1 hr in downtime per 24 hr of operating time would easily make up the loss, without counting the reduced maintenance cost.

Thus, by analyzing various flowsheet arrangements, an alternate process configuration that obtained 52 percent oil yield compared with 56.8 percent for the base case while eliminating maintenance and operating problems of the base case was achieved. It should be noted that while the computer simulation consists of eight or nine calculational blocks, these blocks consisted of three basic unit operations—mixers, splitters, and controllers. This process was simulated without any computer programming requirement using equipment unit operation models that had been developed for the pulp and paper industries. Once the base case was simulated and compared with existing process operating data, the process engineer could rapidly evaluate alternative flowsheet configurations with only minor input modifications to the data file. Steady-state material balance calculations were performed for an iterative recycle problem. If done by hand, these calculations would be time-consuming and tedious, but they were carried out in a matter of seconds on the computer. Thus, the process engineer has more time available for devising new configurations and analyzing their performance in order to develop improved processes.

Potato Blanching

In the French fry or Par-fry process, potatoes are blanched to remove sugar for product color control during subsequent frying. A typical potato blancher system configuration is shown in figure 5.21. The blanching op-

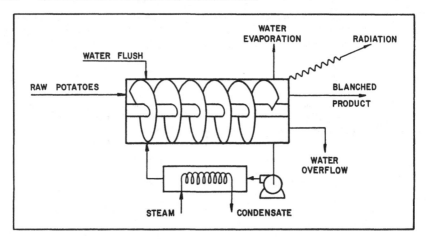

FIG. 5.21.   Potato blancher system configuration. (From Drown and Petersen, 1983.)

eration consists of submerging raw potatoes in hot water, which is recirculated through a heat exchanger to provide the energy to heat the raw potatoes from ambient conditions to blanch temperature. Sugar is leached from the potatoes into the blanch water. Water flush is added to the blancher and overflows, controlling sugar content in the system. Water is also lost by evaporation from the top of the vessel. There are heat losses by radiation and natural convection to the surroundings. The computational potato blancher block diagram, shown in figure 5.22, consists of two blocks, one for the leaching blancher and one for the heat exchanger.

Owing to the large water recirculation rates relative to the throughput rates in the system, the blancher can be modeled as a continuous stirred tank reactor, the rate-controlling step for leaching sugar from the potato being diffusion within the potato slab. The solute concentration of the blanch water from the slab is given by Fick's law of diffusion. In addition, the transient conduction of heat within the potato pieces can be analyzed. Evaluation of the centerline temperature of the potato is necessary to ensure that adequate blanching and enzyme degradation have been achieved. The differential equations for mass and energy have been integrated analytically and solved to predict the outlet concentrations and energy requirements of the blancher. In addition to sugar, leaching of minerals and vitamins can be predicted with the same model provided that the inlet concentrations and the effective diffusion coefficients are known.

In analyzing the energy inputs and outputs of the blancher, water evaporation, while accounting for a very small water mass loss, was shown to represent a significant energy requirement. Appreciable energy savings can be achieved in commercial blanchers by minimizing the surface area of hot water exposed to the ambient atmosphere. By using the predictive leaching model it is possible to predict the effects of residence time, op-

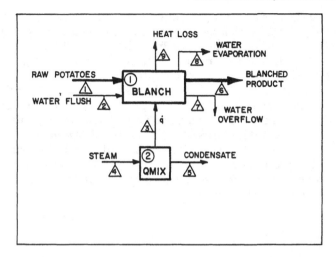

FIG. 5.22.   Potato blancher GEMS block diagram. (From Drown and Petersen, 1983.)

erating temperature, and water flush rates on product sugar content and energy consumption in order to achieve optimum operating conditions and thereby produce a desired product with minimum energy input, as shown by Drown and Petersen (1983).

This section has shown that computer simulation helps process engineers to identify and evaluate many alternative methods of operating existing equipment and the design engineers to evaluate multiple design options. This can result in significant savings in both capital and operating costs. Utilizing a computerized flowsheet material and energy balance simulation as a calculational tool allows process engineers to concentrate their efforts on devising alternatives and evaluating their cost effectiveness and on understanding the chemical and physical phenomena that are going on within the process, while reducing the drudgery of tedious hand calculations of iterative recycle systems.

TRAINING REQUIREMENTS

In a final note on this chapter, it should be emphasized that although process models are extremely useful to food scientists, their development requires considerable training and expertise in engineering science and mathematics at advanced levels that lie considerably outside the field of food science. Graduate level training is typically required in such areas as engineering thermodynamics, heat and mass transfer, reaction kinetics, and engineering mathematics, including intermediate and partial differential equations, numerical analysis, matrix algebra, and software engineering. Furthermore, the equivalent of a basic undergraduate degree in engineering is an essential prerequisite for such advanced engineering training. Therefore, the successful development of such mathematical process models is most likely to result from the efforts of highly trained engineers with sufficient understanding of food science to recognize and develop applications to food process operations.

The emerging field of food engineering as an academic discipline is specifically designed for this purpose. Through cooperation among agricultural engineering, chemical engineering, and food science departments a number of colleges and universities already offer advanced degree programs in food engineering. The development of mathematical models for computer simulation of food processing operations is only one example of the many ways in which highly trained food engineers are contributing toward advancing technology in the food industry.

REFERENCES

ARCE, J. A., POTLURI, P. L., SCHNEIDER, K. C., SWEAT, V. E., and DUTSON, T. R. 1983. Modeling beef carcass cooling using a finite element technique. Transactions of the ASAE 26(3): 950.

ARMIZER, W. B. 1979. Computer Applications in Fermentation Technology. John Wiley & Sons, New York.

ASSOCIATION OF VITAMIN CHEMISTS. 1951. Methods of Vitamin Assay. Interscience, New York.

BALL, C. O. 1938. Advancement in sterilization methods for canned foods. Food Research 13(3): 13.

BALL, C. O., and OLSON, F. C. W. 1957. Sterilization in Food Technology. McGraw-Hill, New York.

BARRETT, J. R., OKOS, M. R., and STEVENS, J. B. 1981. Simulation of low temperature wheat drying. Transactions of the ASAE 24(4): 1042–1046.

BERTIN, R., PIERRONNE, F., and COMBARNOUS, M. 1980. Modeling and simulating a distributed parameter tunnel drier. Journal of Food Science 45: 122.

BLOORE, C. G., and BOAG, I. F. 1982. A simulation model of a tall-form spray drier. New Zealand Journal of Dairy Science and Technology 17(2): 121–134.

BONACINA, C., and COMINI, G. 1973. On a numerical method for the solution of the unsteady state conduction equation with temperature dependent parameters. Proceedings of the 13th International Congress of Refrigeration 2: 329.

BONACINA, C., COMINI, G., FASANO, A., and PRIMICERIO, M. 1973. Numerical solutions of phase change problems. International Journal of Heat and Mass Transfer 16: 1825.

BONACINA, C., COMINI, G., FASANO, A., and PRIMICERIO, M. 1974. On the estimation of thermal physical properties in nonlinear heat conduction problems. International Journal of Heat and Mass Transfer 17: 861.

CHEN, C. S., CARTER, R. D., MILLER, W. M., and WHEATON, T. A. 1980. Energy performance of a HTST citrus evaporator under digital computer control. ASAE Paper No. 80-6028. American Society of Agricultural Engineers, St. Joseph, MI.

CLELAND, A. C., and EARLE, R. L. 1982. Freezing time prediction for foods—a simplified procedure. International Journal of Refrigeration 5(3): 134.

DROWN, D. C., and PETERSEN, J. N. 1983. Application of flowsheeting in the food process industry. ASAE Paper No. 83-6523. American Society of Agricultural Engineers. St. Joseph, MI.

EDWARDS, L., and BALDUS, R. 1979. GEMS-User's Manual. University of Idaho, Moscow, ID.

FELLICIOTTI, E., and ESSELEN, W. B.   1957.   Thermal destruction rates of thiamine in pureed meats and vegetables. Food Technology 11(2): 77–84.

FRASER, B. M., and MUIR, W. E.   1981.   Airflow requirements predicted for drying grain with ambient and solar-heated air in Canada. Transactions of the ASAE 24(1): 208–210.

HAYAKAWA, K. I.   1964.   Development of formulas for calculating the theoretical temperature history and sterilizing value in a cylindrical can of thermally conductive food during heat processing. Ph.D. dissertation, Rutgers, the State University, New Jersey.

HAYAKAWA, K. I.   1969.   New parameters for calculating mass average sterilizing value to estimate nutrients in thermally conductive food. Canadian Institute of Food Technologists 2(4): 165.

HELDMAN, D. R.   1974.   Computer simulation of food freezing processes. Proceedings of the 6th International Congress of Food Science and Technology 4: 397.

HELDMAN, D. R.   1982.   Food properties during freezing. Food Technology 36(2): 92.

HELDMAN, D. R.   1983.   Factors influencing food freezing rates. Food Technology 37(4): 103–109.

HELDMAN, D. R., and GORBY, D. P.   1975a.   Prediction of thermal conductivity in food. Transactions of the ASAE 18(1): 740.

HELDMAN, D. R., and GORBY, D. P.   1975b.   Computer simulation of individual-quick-freezing of foods. ASAE Paper No. 75-6016. American Society of Agricultural Engineers, St. Joseph, MI.

HOHNER, G. A., and HELDMAN, D. R.   1970.   Computer simulation of freezing rates in foods. Presented at 30th Annual Meeting of the Institute of Food Technologists, 24–27 May, San Francisco.

HOLDREDGE, R. M., and WYSE, R. E.   1982.   Computer simulation of the forced convection cooling of sugar beets. Transactions of the ASAE 25(5): 1425.

HSIEH, R. C., LEREW, L. E., and HELDMAN, D. R.   1977.   Prediction of freezing times for foods as influenced by product properties. Journal of Food Processing Engineering 1: 183.

JEN, Y. Y., MANSON, J. E., STUMBO, C. R., and ZAHRADNIK, J. W.   1971.   A simple method for estimating sterilization and nutrient and organoleptic factor degradation in thermally processed foods. Journal of Food Science 36: 692.

LIMA HON, V. M., CHEN, C. S., and MARSAIOLI, A. Jr.   1979.   Computer simulation of dynamic behavior in vacuum evaporation of tomato paste. Transactions of the ASAE 22(1): 215.

MEIERING, A. G., and SUBDEN, R. E.   1983.   Fermentation control by microcomputers. ASAE Paper No. 83-6534. American Society of Agricultural Engineers, St. Joseph, MI.

NAGAOKA, J., TAKAGI, S., and HOTANI, S.   1955.   Experiments on the freezing of fish in an air-blast freezer. Proceedings of the 9th International Congress of Refrigeration, Paris 2:4.

NAVEH, D., KOPELMAN, I. J., and PFLUG, I. J.   1983.   The finite element method in thermal processing foods. Journal of Food Science 48: 1086.

NEELAKANTAN, P. S., and MUKESH, D.   1979.   Computer model of a continuous evaporative crystallizer. Industrial and Engineering Chemistry, Process Design and Development 18: 56.

PISIPATI, R., and FRICKE, A. L.   1979.   Computer simulation of a single screw cooking extruder. Proceedings of the 2nd International Congress of Engineering and Food (Helsinki), and International Union of Food Science & Technology.

PURWADARIA, H. K., and HELDMAN, D. R.   1982.   A finite element model for prediction of freezing rates in food products with anomalous shapes. Transactions of the ASAE 25(3): 827.

RADOVIC, L. R., TASIC, A. Z., GROZDANIC, D. K., DJORDJEVIC, B. D., and VALENT, V. J.   1979.   Computer design and operation of a multiple-effect evaporator system in the sugar industry. Industrial and Engineering Chemistry, Process Design and Development 18: 318.

SABBAH, M. A., MEYER, G. E., KEENER, H. M., and ROLLER, W. L. 1979. Simulation studies of reversed-direction-air-flow drying method for soybean seed in a fixed bed. Transactions of the ASAE 22(5): 1162–1166.

SAGUY, I., KOPELMAN, I. J., and MIZRAHI, S. 1978a. Thermal kinetic degradation of betanine and betalanic acid. Journal of Agricultural and Food Chemistry 26(2): 360.

SAGUY, I., KOPELMAN, I. J., and MIZRAHI, S. 1978b. Computer-aided determination of beet pigments. Journal of Food Science 43: 124.

SAGUY, I., KOPELMAN, I. J., and MIZRAHI, S. 1980. Computer-aided prediction of beet pigment (betanine and vulgaxanthin-I) retention during air-drying. Journal of Food Science 45: 230–235.

STUMBO, C. R. 1965. Thermobacteriology in Food Processing. Academic Press, New York.

TEIXEIRA, A. A. 1971. Thermal process optimization through computer simulation of variable boundary control and container geometry. Ph.D. dissertation, University of Massachusetts, Amherst.

TEIXEIRA, A. A. 1978. Conduction-heating considerations in thermal processing of canned foods. Paper No. 78-WA/HT-55, American Society of Mechanical Engineers, United Engineering Center, New York.

TEIXEIRA, A. A., DIXON, J. R., ZAHRADNIK, J. W., and ZINSMEISTER, G. E. 1969a. Computer optimization of nutrient retention in the thermal processing of conduction-heated foods. Food Technology 23(6): 137.

TEIXEIRA, A. A., DIXON, J. R., ZAHRADNIK, J. W., and ZINSMEISTER, G. E. 1969b. Computer determination of spore survival distributions in thermally processed conduction-heated foods. Food Technology 23(3): 78.

TEIXEIRA, A. A., STUMBO, C. R., and ZAHRADNIK, J. W. 1975a. Experimental evaluation-of mathematical and computer models for thermal process evaluation. Journal of Food Science 40: 653.

TEIXEIRA, A. A., ZINSMEISTER, G. E., and ZAHRADNIK, J. W. 1975b. Computer simulation of variable retort control and container geometry as possible means for improving thiamine retention in thermally processed foods. Journal of Food Science 40: 656.

THOMPSON, J. E., KRANZLER, G. A., and STONE, M. L. 1982. Microcomputer control for hop drying. ASAE Paper No. 82-5520. American Society of Agricultural Engineers, St. Joseph, MI.

TROEGER, J. M., and BUTLER, J. L. 1979. Simulation of solar peanut drying. Transactions of the ASAE 22(4): 906.

6

Process Optimization

ELEMENTS OF OPTIMIZATION THEORY

Once a mathematical model has been established, optimization of the process, that is, finding the unique set of process conditions that produces the best results, is carried out. An example of process optimization in the thermal processing of canned foods would be to find the combination of process time and retort temperature that produces optimum product quality while achieving the necessary level of sterilization.

The following five elements are common to all optimization problems (Evans, 1982):

Performance function: This is the quantity to be maximized (or minimized); it is often referred to as the profit, the cost, or the *objective function.* In the thermal process example the objective function could be to minimize thiamine degradation (or maximize thiamine retention) as an indication of product quality. Whether measured in dollars, efficiency, or thiamine content, the performance function simply evaluates the objective functions for any given set of process conditions or variables.

Decision variables: These are the parameters in the process that can be adjusted to improve performance. They are the free and independent variables that must be specified in the process model, such as the retort temperature and the process time in thermal processing for canned food sterilization.

Constraints: All optimization problems have constraints on the allowed solutions. These constraints may limit values of the decision variables, other dependent variables, or even other performance functions of the process. In the thermal processing example, constraints on the decision variables would be upper and lower limits for the retort temperature imposed by limitations of equipment operation. A more critical constraint would be the requirement that a specified level of sterilization or bacterial lethality be achieved in all cases (i.e., a limiting value placed on another performance function of the same process), thus constraining the degree to which thiamine retention could be maximized.

Mathematical model: As emphasized earlier, the need for a mathematical model to simulate the process is fundamental to any reasonable attempt at process optimization. The only other alternative is trial-and-error physical experimentation through costly laboratory or pilot-plant experiments. The model is the mathematical representation of the process that determines the performance functions in terms of the decision variables or other independent variables subject to constraint. When programmed for computer simulation, the model can predict process results for any one set of decision variables in a matter of a few minutes at essentially negligible cost.

Optimization technique: This is simply the method by which all possible combinations of decision variables are searched within the boundary of the constraints in order to ensure that the optimum combination is indeed found. For problems in which the decision variables are constrained to a limited number of combinations, the optimization technique can be as simple as methodically searching all possible combinations to select the one that produces the best results from the objective functions in the mathematical model. For more complex problems with many decision variables that can take on a wide range of values, the number of search possibilities can become almost limitless. In such situations it is often necessary to resort to a structured technique based on some form of mathematical algorithm that is suitable for the type of problem to be solved.

The relationship between the elements of an optimization problem and their role in its solution is illustrated in figure 6.1. The case study examples covered in this chapter will focus on the first four of the elements discussed above. The study of optimization techniques is mathematically involved and will not be considered here. However, considerable work describing the application of optimization techniques to a number of food processing

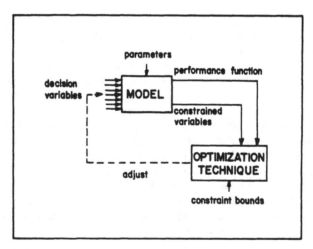

FIG. 6.1. Elements of an optimization problem. (From Evans, 1982.)

operations has recently appeared in the literature. Karel et al. (1984) have prepared a critical review of optimization methods and available software for food engineering applications. Saguy and Karel (1979) have described the use of Pontryagin's maximum principle to the case study example on thermal processing of canned foods covered later in this chapter. Application of two different optimization techniques in food dehydration is described by Mishkin et al. (1982), and a comprehensive treatment on computer-aided optimization techniques for food process applications is given by Saguy (1983).

The case study examples in this chapter will make use of the kinetic models and process models developed in Chapter 5 for the thermal processing of canned foods. By applying the elements of optimization theory discussed above, optimization problems of varying complexity will be formulated and solved to illustrate the powerful use that can be made of process models in process optimization once the models are available.

STEADY-STATE OPTIMIZATION IN THERMAL PROCESSING

The preceding section has already suggested that a classic optimization problem for thermally processed canned foods would be to find the combination of process time and retort temperature that would maximize product quality (thiamine retention) while ensuring required sterilization. The retort temperature, once chosen, will remain constant over the process time, which means that control of the process (and, thus, the optimization problem) is steady-state. Steady-state optimization differs from dynamic optimization, in which the retort temperature is allowed to vary as a function of time; the latter situation will be treated later in this chapter.

Kinetic Models as Basis for Optimization

Since time and temperature affect both the inactivation of bacteria and the degradation of thiamine in the same product, the time- and temperature-dependent kinetics for each of these factors must be different sufficiently from each other to permit time-temperature combinations that are equivalent with respect to bacterial inactivation to produce differing degrees of thiamine degradation. Otherwise, both factors will remain in constant proportion to each other for all process conditions and no optimization will be possible.

Both the inactivation rate of bacterial spores and the degradation rate of thiamine accelerate with increasing temperature, as shown by the kinetic models developed in Chapter 5. The important basis for optimization, however, is that this accelerating effect is much greater on the bacterial kinetics than on the thiamine kinetics. The significance of this difference is dramatically illustrated in figure 6.2 (from Joslyn and Heid, 1963). In this figure a family of thermal retention versus time curves representing different levels of thiamine degradation are shown superimposed on one thermal death versus time curve representing a 90 percent reduction (one logarithmic cycle) in bacterial spore survival. Since the figure is a semilogarithmic plot of time (decimal reduction time) versus temperature, all the points on any one line represent combinations of time and temperature that have equivalent effects in achieving the level of destruction represented by that line.

Assuming that the thermal death versus time curve for 90 percent reduction of bacterial spores in figure 6.2 is a process constraint on the optimization problem, then only time-temperature combinations that fall on this curve can be allowed. Since this curve intersects with curves representing lower levels of thiamine destruction as conditions of higher temperature with correspondingly shorter times are approached, then by observation the optimum process is that occurring at the highest temperature possible with the associated minimum process time.

The benefits of high temperature–short time processing have been known for a number of years and have led to the widespread use of aseptic packaging methods wherever possible. These methods generally apply only to fluid products that can be pumped through heat exchangers capable of applying ultrahigh temperature–short time processing conditions to the product before it is filled and sealed aseptically into previously sterilized packages. Table 6.1, from Everson et al. (1964), compares the thiamine retention obtained from a retort process with that obtained from an aseptic process of equal sterilizing value for strained lima beans and strained beef.

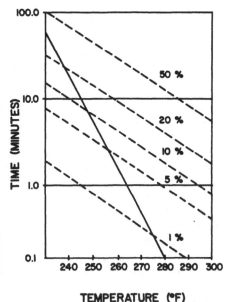

FIG. 6.2. Semilogarithmic plot showing thiamine degradation curves (dashed lines) superimposed on a thermal death versus time curve (solid line) for bacterial spores. (From Joslyn and Heid, 1963.)

TABLE 6.1. Comparison of Thiamine Retention from Equivalent Retort and Aseptic Thermal Processes on Strained Lima Beans and Strained Beef

Product	Processing Method	Thiamine Retention %
Strained lima beans	Retort process (53 min at 240°F)	57.8
	Aseptic process (13 s at 300°F)	85.6
Strained beef	Retort process (48 min at 250°F)	80.1
	Aseptic process (1 min at 300°F)	92.8

(*Source:* Everson et al., 1964.)

This review of the kinetic models has shown that thiamine retention can be improved significantly if the product can be heated and cooled rapidly to actually experience high temperature–short time processing. In a retort process with conduction-heated canned foods, the use of higher retort temperatures with shorter times may not necessarily provide benefit. Regions near the surface of the container must be exposed to the higher temperature sufficiently long to permit adequate sterilization at the can center. The accelerating effect of the higher temperature on the rate of thiamine degradation in these regions over this length of time is such that the overall saving in process time may not be sufficient to compensate for such loss. At the other extreme, the use of lower retort temperatures with longer process times would also produce poor results because of the relative kinetics discussed previously. Therefore, some optimum combination should exist for any given product by which maximum thiamine retention can be achieved. This forms the basis for the steady-state optimization problem to be formulated in this case study example.

Formulating the Problem

The first element to be addressed in formulating the optimization problem is the objective function, which for this case study example would be to maximize thiamine retention. Mathematically, the objective function would be evaluated by the computer model developed in Chapter 5 using kinetic data for the thermal degradation of thiamine. The total integrated thiamine content in the product reported at the end of the simulated process would be the "result" of the objective function from each choice of decision variables. The model itself is the key element that makes the optimization problem practical.

The last two elements to be considered are the decision variables and their constraints. The sterilization constraint requires that every combination of time and temperature must result in a specified degree of bacterial lethality. This constraint has far-reaching implications on the choice of decision variables. For any given retort temperature only one process time will satisfy this constraint. Thus, the process time is not a decision variable at all; it is a dependent variable dictated by the sterilization constraint for each retort temperature chosen. Therefore, the optimization problem is reduced to the simple case of a single decision variable (retort temperature) and objective function (thiamine retention). In this form the problem lends itself readily to the simplest optimization technique, a direct search over a limited range of retort temperatures. The range of retort temperature to be searched is limited on the high side by equipment capability and on the low side by temperatures below which there is no lethal effect on bacterial spores.

Search for Optimum Process

The optimization technique used to solve this problem was reported by Teixeira et al. (1969) and involved a two-step approach. In the first step the computer model was used to determine the process time that would be required with each retort temperature to achieve the specified level of bacterial lethality to comply with the sterilization constraint imposed on the problem. This step resulted in a number of retort temperature and process time combinations that were equivalent in their ability to achieve the required degree of sterilization. A curve describing the loci of these equivalent processes is shown in figure 6.3, where retort temperature has been plotted against process time.

Any point on this curve represents a thermal process that produces the required degree of sterilization for the given product. In the second step the computer model was used to predict the thiamine retention associated

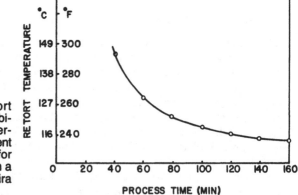

FIG. 6.3. Loci of retort temperature–time combinations representing thermal processes of equivalent sterilization capability for conduction-heated food in a No. 2 can. (From Teixeira et al., 1969.)

with a number of these processes by evaluating the objective function for selected values of the decision variable. These thiamine retention values were plotted against process time with the corresponding retort temperatures in figure 6.4. The resulting curve clearly shows that thiamine retention reaches a maximum within the range of search, revealing the optimum process time to be 90 min at a retort temperature of 120°C.

Note that the optimum process, contrary to expectations, does not favor the high temperature–short time combinations. In fact, the shape of the curve indicates that thiamine retention decreases rapidly as such processes are approached, compared with the rate of decrease when the process goes to lower temperature and longer time. This behavior is a result of the integrated effect of the nonuniform temperature distributions that exist throughout the container during heating and cooling. This integrated effect cannot be accounted for by comparing the rate data, because the rate data for both thiamine and bacteria are derived from laboratory conditions of uniform temperature.

If other nutrients of quality factors were found to be heat-sensitive in the same way as thiamine (i.e., to have the time and temperature dependence of a first-order degradation reaction), and if their rate data were known, then their retentions could be plotted as in figure 6.4. By defining a criterion function that correlates the importance of each factor, an optimum process could be obtained by subjecting the retention data to the criterion function and finding its maximum value for the equivalent processes. Such a situation is illustrated by figure 6.5, where retentions have

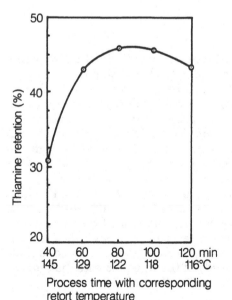

FIG. 6.4. Optimization curve showing percent thiamine retention over range of thermal processes, designated by process time with corresponding retort temperature, that are equivalent with respect to sterilization capability. (From Teixeira et al., 1969.)

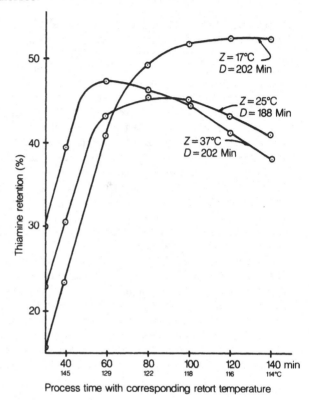

FIG. 6.5. Multinutrient optimization showing percent nutrient retention versus sterilization-equivalent thermal processes for nutrients with different kinetic rate constants. (From Teixeira et al., 1969.)

been plotted for two other, hypothetical nutrients differing from thiamine and from each other in their degradation rate characteristics.

Since heat-sensitive nutrients can be distinguished from each other by their rate data expressed in terms of a D and Z value, just as in thermobacteriology, it was interesting to note what effect these values had on the optimization curve. As can be seen from figure 6.5, a nutrient with a Z value lower than that of thiamine indicates an optimum to the right favoring a long time–low temperature process, whereas the high Z-valued nutrient favors a high temperature–short time process. A change in D value, simply shifts the curves up or down, as shown in figure 6.6, in which each curve was constructed with the same Z value but different D values. Physically, these curves could represent the retention of thiamine in different foods receiving the same heat treatment. Although these multinutrient considerations are hypothetical, they have been included to show the possible advantages of further study of the rate processes that govern the many changes occurring in foods during thermal processing.

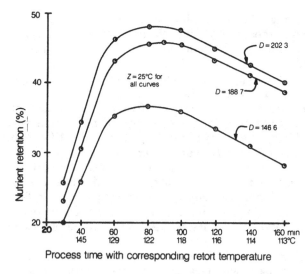

FIG. 6.6. Optimization for thiamine retention in different products (designated by different D values) over range of sterilization-equivalent thermal processes. (From Teixeira et al., 1969.)

DYNAMIC OPTIMIZATION IN THERMAL PROCESSING

The following case study example addresses the same optimization problem that was presented in the preceding section, with the one important difference that the decision variable (retort temperature) will be allowed to vary as a function of time. The problem objective will be to find the optimum function of time (retort temperature–time policy) that produces a maximum level of thiamine retention while satisfying the constraint on required sterilization. Since the decision variable is a function of time, the problem becomes one of dynamic optimization.

Problem Rationalization

A review of kinetic models developed in Chapter 5 has shown that improved thiamine retention can be achieved with a high temperature–short time policy when the material can be heated rapidly. The shorter process time permitted by the accelerated lethal rate prevents the thiamine degradation from proceeding very far even though this degradation occurs at a faster rate.

It has also been shown, however, that when high temperature–short time policies have been applied to conduction-heated canned foods, the resulting thiamine retention is lower (Teixeira et al., 1969) as a consequence of the physical size and thermal properties of the material being heated. Because regions near the surface of the container must be exposed to the higher temperature sufficiently long to permit adequate sterilization at the center, the accelerating effect of the higher temperature on the rate of thiamine degradation in these regions is such that the reduction in process time is not sufficient to compensate for it. The rationale for dynamic optimization was based on the hypothesis that it is possible to adjust the degree to which each of these opposing effects predominates by manipulating the surface temperature as a function of time.

This problem is formulated by addressing the five elements of optimization theory presented at the beginning of this chapter in the same way as the steady-state problem. The exciting challenge lies in identifying an optimization technique to effectively search over the virtually infinite number of temperature-time policy possibilities. Mathematical techniques such as Pontryagin's maximum principle (Pontryagin et al., 1962) are applicable to this type of problem. The Pontryagin technique was, in fact, used by Saguy and Karel (1979) to solve the very problem presented in this case study example, and the results of their work will be presented later in this chapter.

Attempts to use such mathematical techniques are often discouraging to food scientists and engineers who lack the rigorous training required to understand the involved mathematics and algorithms. The purpose of this case study example is to illustrate how a reasonably good solution to this type of problem can be obtained without resorting to such elegant mathematical techniques. Although the approach is direct search by trial and error, the reader can appreciate how effective such a simple technique can be with the availability of a good predictive process model and sufficient knowledge of practical limitations on the process.

Search by Trial and Error

The use of direct search by trial and error to solve this optimization problem was described by Teixeira et al. (1975), and was the first reported attempt at dynamic optimization for the thermal processing of canned foods. The key to the successful use of this approach is the ability to narrow down the number of search possibilities to a reasonable few through judicious application of practical process considerations (i.e., constraints on the process), as described below.

In order to limit the scope of the investigation, the assumption has been made that any practical application of variable surface temperatures in the thermal processing of canned foods would be achieved with the use of saturated steam under pressure, either as a variable retort control or by passing the food container from one chamber to another held at a different steam temperature. Because of the low thermal diffusivity and the container size of most canned foods, the time of exposure to any given surface temperature would have to be of at least several minutes duration in order to affect the interior temperature distribution. With exposure times of this magnitude, it was reasoned that surface temperatures above 265°F should be avoided to prevent excessive degradation in the outer regions of the container. The lower limit for the surface temperature was taken as 225°F because at lower temperatures there would be little or no lethal effect on thermophilic spores while nutrient degradation would continue at a significant rate.

Because of its widespread use, the container chosen for this study was a No. 2 can, which measured $3\frac{7}{16}$ inches in diameter by $4\frac{9}{16}$ inches in height. The thermal diffusivity of the food material that was modeled in this example was taken as 0.0143 in²/min, which has been found to be representative of most conduction-heated foods.

Spores of *Bacillus stearothermophilus* were taken as the food spoilage organism of greatest concern. The destruction rate data for these organisms, as reported by Stumbo (1965), were given by a D value of 4 min at 250°F and a Z value of 18°F. The sterilizing effect of all the processes defined in this investigation was held fixed as a constraint and was defined by a five-logarithmic-cycle reduction of the initial spore population in the container.

According to Felliciotti and Esselen (1957), thiamine in pork has a D value at 246°F of 178.6 min and a Z value of 46°F. These values were chosen for the thiamine destruction rate data used in this example.

In order to establish a definition of *improved thiamine retention*, a standard thermal process was defined, and the thiamine retention for this process was accepted as a reference standard. This *standard process* was described as corresponding to the process time required to achieve a five-logarithmic-cycle reduction in spores of *B. stearothermophilus* in a No. 2 can with a thermal diffusivity of 0.0143 in²/min at a constant surface temperature of 250°F. A hot-fill initial temperature of 160°F was assumed for all processes throughout the investigation.

Theoretically, there were an infinite number of ways in which the surface temperature could vary between the upper and lower limits specified. However, three general types of behavior functions were chosen to classify nearly all the possibilities, namely, sinusoidal functions, ramp functions, and step functions. The method of search was to systematically investigate surface temperature policies in each of these categories.

The purpose of investigating sinusoidal policies was to determine whether or not there would be some benefit to a policy in which the surface temperature reversed itself periodically between upper and lower limits. The amplitude was fixed so that the surface temperature oscillated between 265 and 225°F at various frequencies. The lower limit for the range of frequencies studied was determined by noting the process time for the standard process and choosing a frequency that would permit half of a sine wave within that time. The upper limit was established by noting that frequency above which the results were no different than when the surface temperature was simply held constant at the mean value of the sine function. The investigation began with a frequency of one-half cycle per hour and progressed to one cycle per hour, two cycles per hour, and so on. This same sequence was then repeated with a 90° shift in phase angle to investigate a corresponding series of cosine functions. The process for each policy was determined by repeated computer calculations in a search routine to find the time at which cooling should begin to achieve the required five-logarithmic-cycle reduction in spore population at the end of cooling.

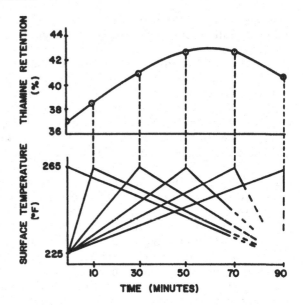

FIG. 6.7. Results from computer search of linear ramp functions, showing percent thiamine retention versus individual ramp policies of surface temperature as a function of time. (From Teixeira et al., 1975.)

A preliminary investigation of possible ramp functions for the surface temperature was restricted to rising and falling ramps of different but constant slopes, in which the surface temperature increased linearly from the lower limit to the upper limit and then decreased linearly to the point at which cooling should begin for the required sterilization. The sequence of investigation consisted of locating the apex of the ramp functions at different points in time within the process time, as shown in the lower portion of figure 6.7. Once the optimum ramp function had been found from this investigation, attempts were made to obtain further improvement by investigating nonlinear ramp functions such as exponentials and quadratics in the neighborhood of this "best" ramp.

TABLE 6.2. Maximum Thiamine Retention Observed from Investigations of Various Surface Temperature Policies, Including Process Times Required to Ensure Equal Sterilization

Surface Temperature Policy (Fig. 6.8)	Process Time (min)	Thiamine Retention (%)
Standard process (a)	89	41
Best sinusoidal function (b)	70	41
Best combination of ramps (c)	88	43
Best single square wave (d)	79	41
Best sequence of steps (e)	84	43

(*Source:* Teixeira et al., 1975.)

Additional investigations with step functions were based on information obtained from the results of the previous investigations with sinusoidal and ramp functions. Successive trials with step functions were made on the basis of information obtained from all previous results.

A synopsis of the results from the investigation of variable surface temperature policies is presented in Table 6.2, which shows the maximum thiamine retention observed with the optimal policy in each category, along with the process time required in each case to ensure equal sterilization for all policies. Figure 6.8 has been included to present a graphical description for each of these policies. It should be noted that the policies producing the highest level of thiamine retention seem to favor a general configuration in which the surface temperature gradually increases from the lower limit to reach the upper limit approximately midway through the process time and then gradually decreases to the point at which cooling should begin. Figure 6.7, showing the results from the investigation of linear ramp functions, clearly reveals this trend.

The results show that the maximum thiamine retention observed was only two percentage points greater than that for the standard process, representing only 5 percent improvement. Since Felliciotti et al. (1956) reported deviations in their laboratory assay of thiamine within ± 5 percent of the mean, the maximum thiamine retention observed from this investigation cannot be accepted as any indication of significant improvement.

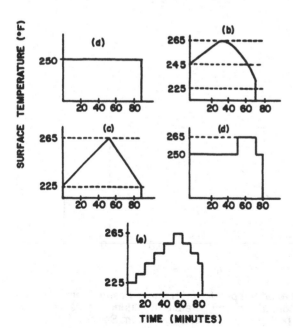

FIG. 6.8.   Variable surface temperature–time profiles of equivalent sterilization capability that produced maximum thiamine retention in each category. (From Teixeira et al., 1975.)

Search by Maximum Principle

As mentioned earlier, Saguy and Karel (1979) approached this problem using a mathematical algorithm, (Pontryagin's maximum principle) for the optimization technique as a more elegant alternative to the trial-and-error search of Teixeira et al. (1975). Although no attempt will be made to explain or describe the use of this optimization technique, the results obtained by Saguy and Karel (1979) will be shown for comparison with those obtained by Teixeira et al. (1975).

One of the advantages of the maximum principle when applied to this problem is that it searches out the optimum surface temperature policy as a smooth, continuous function of time, which truly maximizes the objective function if a maximum does indeed exist. Figure 6.9 shows the optimum retort temperature profile to maximize thiamine retention in a No. 2 can of pork puree, as reported by Saguy and Karel (1979). Optimum retort temperature profiles for pea puree in a No. 303 can and for pork puree in a No. 2½ can are shown in figure 6.10, also from Saguy and Karel (1979).

It is interesting to note that in all cases the optimum profile conformed to a general configuration in which the surface temperature gradually increases from the lower limit to reach the upper limit approximately midway through the process time and then gradually decrease to the point at which cooling should begin. This is precisely the same configuration reported for the optimum surface temperature policy by Teixeira et al. (1975) using the trial-and-error technique described previously.

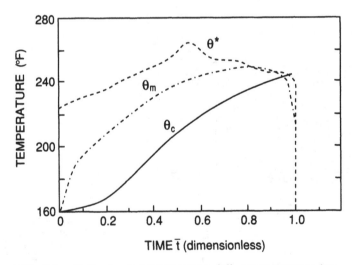

FIG. 6.9.   Optimal retort temperature (θ^*), mass average temperature (θ_m), and central point temperature (θ_c) during the sterilization process of pork puree in a No. 2 can. (From Saguy and Karel, 1979.)

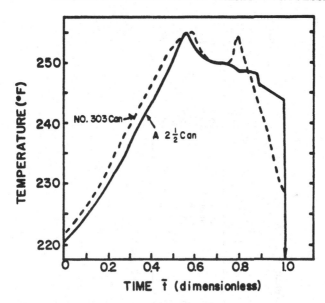

FIG. 6.10. Optimal retort temperature profiles for the sterilization of pea puree in a No. 303 can and pork puree in a No. 2½ can. (From Saguy and Karel, 1979.)

The optimum temperature profiles found by the maximum principle produced slightly higher levels of thiamine retention than the profiles found by trial-and-error, as shown in Table 6.3. Considering that Teixeira et al. (1975) had already optimized the process, the fact that any further improvement was introduced by using the maximum principle is very significant. Moreover, the maximum thiamine retention found and the corresponding temperature profiles that were computed provided a single solution to the problem. It may thus be concluded that a best single process to optimize nutrient retention does indeed exist.

These have been case study examples presented to illustrate how both simple (trial-and-error) and elegant (maximum principle) optimization techniques can be used to solve a dynamic optimization problem in the thermal processing of canned foods. The implications of this work for industry practice, however, are not likely to be significant. For one thing, the industry is required by federal regulations (FDA Low-Acid Canned Food Regulations) to maintain careful control of retort temperature, with costly exceptions for any deviation from the specified temperature. Allowing the retort temperature to vary with time would pose an unwelcome complication to such regulations. Moreover, the results of this work have shown that the difference between the maximum thiamine retention possible with variable surface temperature and the normal level of thiamine retention obtained with constant retort temperature has no commercial significance. Improvement in thiamine retention by such means is physically limited

TABLE 6.3. Comparison of Methods for Optimizing Thiamine Retention During Heating in Cans

Product	Process time (min)	Can Dimension (in)	Name	Thermal Diffusivity (in²/min)	Retort Temperature (°F)	D 250°F Thiamine (min)	Thiamine Retention (%)
[a]Pork puree	89	307 x 409	A/2	0.0143	250	178.6	41.0
[b]Pork puree	84	307 x 409	A/2	0.0143	Variable*	178.6	43.0
[c]Pork puree	89	307 x 409	A/2	0.0143	Variable[†]	178.6	45.3
[d]Pea puree	85	303 x 406	No. 303	0.0158	250	165.6	49.2
[e]Pea puree	87	303 x 406	No. 303	0.0158	Variable*	165.6	50.2
[f]Pea puree	87	303 x 406	No. 303	0.0158	Variable[†]	165.6	52.4

*Retort temperature based on sequence of steps (see reference).
[†]Retort temperature determined by the optimal temperature profile; see figures 6.9 and 6.10.
[a]Source: Teixeira et al. (1975).
[b]Source: Teixeira et al. (1975).
[c]Source: Saguy & Karel (1979).
[d]Source: Teixeira et al. (1975).
[e]Source: Teixeira et al. (1975).
[f]Source: Saguy & Karel (1979).

by the slow rate of heat conduction in foods and the size and geometry of the cans most commonly used.

In fact, the container geometry is a strong factor that would limit the response of interior temperatures to any control action on the surface. Another useful application of the computer model is to study the effects of various container geometrics of equal volume on the level of thiamine retention for both constant and time-varying surface temperature policies of equal sterilizing value. A case study example describing this application is presented in the next section.

OPTIMUM CONTAINER GEOMETRY IN THERMAL PROCESSING

From what is understood of the relative reaction kinetics involved, it could be reasoned that maximum thiamine retention would favor a temperature policy such that every point throughout the container would experience a high temperature–short time process. However, since the driving force required to raise the temperature of the food must occur at the boundary, the temperature at any interior point can respond only in accord with the mechanics of heat conduction. The rate of this response is severely limited by the depth of food material to which the heat energy must penetrate and the low thermal diffusivity characteristic of most foods.

Although thermal diffusivity is a physical constant describing the thermal properties of the food, the depth of food material through which the heat energy must penetrate can be treated as a control variable by adjusting the container geometry. Thus, taking container geometry as the decision variable, a simple direct search technique can be used to find the optimum container geometry that maximizes thiamine retention as long as reasonable constraints are imposed on the range of geometric configurations to be searched. This is accomplished by judicious application of practical considerations to the problem, as discussed by Teixeira et al. (1975).

In order to be consistent with practical considerations of can filling and sealing methods and to accommodate the provision for a cylindrical geometry in the computer model, the investigation of container geometries was restricted to a right circular cylinder of various height-to-diameter ratios enclosing a constant volume equal to that of a No. 2 can. The height-to-diameter ratio L/D was chosen to range between a low value of 0.1 corresponding to a container height of 0.75 in and a high value of 13.7 corresponding to a container height of 20 in.

The surface temperature was held constant at 250°F to permit direct comparison of thiamine retentions with those of the standard thermal process previously defined. The process time for each case was, again, determined on the basis of achieving the required sterilization at the end of cooling. Beginning with the dimensions of a No. 2 can, process determinations were made for lower values of L/D corresponding to decreasing container heights of 3, 2, 1, and 0.75 in. Higher values of L/D were chosen corresponding to decreasing container radii of 1.5, 1.25, 1.00, and 0.75 in.

In this manner it was possible to calculate the thiamine retention for thermal processes of equal sterilizing value in containers of each of the geometries specified and to report an optimum geometry among these for which the thiamine retention was greatest. Additional improvement in thiamine retention was then found by applying higher constant surface temperatures with correspondingly shorter process times to this optimum geometry. The results of this work are presented graphically in figure 6.11 as percent thiamine retention versus L/D on a logarithmic scale. Note that the standard-size No. 2 can is represented by one of the lowest points on the curve. The corresponding container dimensions, surface area, and process times are listed in Table 6.4.

The results show that for an L/D of 0.1 (flat disk), the level of thiamine retention had increased from 41 to 68 percent with a decrease in the process time required from 90 to 30 min. Improvement of the same order of magnitude was also observed at the opposite extreme with L/D greater than 10. These results were obtained with thermal processes in which the retort temperature was held constant at 250°F.

The flat disk geometry ($L/D = 0.1$) was used in investigating various surface temperature policies, with the results shown in Table 6.5. Although an appreciable increase in the thiamine retention was observed with the linear ramp function, maximum retention was obtained with the high temperature–short time policy. This can be explained by the fact that the improved thiamine retention was a result of the shorter process time made possible by the accelerating effect of higher temperature on the bacterial kinetics. Since the system geometry in this case would permit rapid heating,

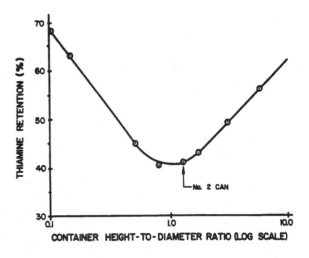

FIG. 6.11. Results from computer search for optimum container geometry showing percent thiamine retention versus height-to-diameter ratio L/D for cylindrical shapes of equivalent volume in which thermal processes of equivalent sterilization capability are carried out. (From Teixeira et al., 1975.)

TABLE 6.4. Container Dimensions, Surface Area, and Process Time for Equivalent Sterilization and Percent Thiamine Retention Corresponding to Various L/D Ratios for Cylindrical Shapes of Equal Volume

L/D	L (in)	D (in)	Surface Area (in²)	Process Time (min)	Thiamine Retention (%)
0.096	0.75	7.8	113.9	30	68
0.143	1.00	7.5	98.5	38	63
0.495	2.00	4.8	66.4	75	45
0.767	3.00	3.9	61.1	90	40
1.270	4.20	3.3	60.6	89	41
1.710	5.12	3.0	62.4	83	43
2.960	7.40	2.5	67.9	68	49
5.760	11.50	2.0	78.5	52	56
13.750	20.60	1.5	100.6	38	63

(*Source:* Teixeira et al., 1975.)

it follows that little would be gained from a variable surface temperature policy other than that in which the surface temperature is held constant at the upper limit.

In addition to the quality improvement that could be expected with these various geometries, there may also be certain economic advantages from the reduced process times required, as shown in Table 6.4. These advantages, however, might be offset by increased container costs relative to the greater surface area required by the various geometries. More importantly, these results have shown that variations in container geometry that permit more rapid heating of the product can be very effective in achieving significant improvement in the nutritional value of thermally processed foods. These considerations would tend to support further development in the use of flexible packaging in thermal processing. Some of the new concepts in flexible packaging, such as the retort pouch made of laminated plastics and foil, lend themselves readily to geometric configurations consisting of thin cross sections that would promote rapid heat transfer.

TABLE 6.5. Effect of Various Surface Temperature Policies on a Flat Disk Geometry with L/D = 0.1

Policy Description*	Thiamine Retention (%)[†]
Standard: $T = 250°F$, $t = 30$ min	68
HT-ST: $T = 265°F$, $t = 13.5$ min	77
Ramp: $T = 265-225°F$, $t = 27$ min	74

*T = retort temperature; t = process time; HT-ST = high temperature–short time.
[†]Same product in standard No. 2 can requires 90-min process time at 250°F with only 41 percent thiamine retention.
(*Source:* Teixeira et al., 1975.)

With regard to the adaptation of metal containers to these geometries, it could be expected that fabrication costs would increase because of the greater surface area requirements, but these could be offset by the economic advantages of reduced process time. A complete system cost analysis would be required to make any conclusive statements in this regard.

OTHER APPLICATIONS OF OPTIMIZATION TO HEAT PROCESSING

In addition to Teixeira et al. (1975) and Saguy and Karel (1979), whose studies have been described in these case study examples, other investigators have recently contributed to optimization in heat processing. Thijssen et al. (1978) developed a shortcut method for calculating sterilization conditions for optimum quality retention. Their process model was based on the analytical solution to the partial differential heat conduction equation coupled to the kinetic models through dimensionless parameters.

Norback (1980) reviewed application of several optimization techniques for determining optimal retort control, including dynamic programming, which has not yet been successfully applied to this problem. It is interesting to note that Teixeira (1971) attempted to use dynamic programming. He concluded that it was nonapplicable to conduction-heated canned foods because interior temperatures in discrete regions of the can could not be controlled independently of the surface temperature or of each other. Such discrete-region distributed parameter control is a requirement for application of dynamic programming, which is most applicable to processes involving multistage units, such as those used in grain drying (Brook and Bakker-Arkema, 1978) and in food dehydration (Mishkin et al., 1982).

Martens (1980) also investigated optimal retort control for the thermal processing of flexible retort pouches. Since the geometric configuration of these packages can be represented mathematically as an infinite flat slab for heat transfer purposes, only one-dimensional heat conduction needs to be considered, and the mathematical model for the process is much simpler than that for a finite cylinder. The objective function in Martens' (1980) optimization problem was to maximize the retention of methionine in a conduction-heated product.

Another study on optimization of heat processes was conducted by Hildebrand (1980). He divided the problem into two parts: a process engineering problem centered around determining the best interior temperature profile in the product, and a control engineering problem of finding the optimum surface temperature profile to achieve this predetermined interior temperature profile. A good commentary on all these various techniques and their applications to optimization in heat processing can be found in Lund (1982). Lund (1977) also reviewed the principles involved in maximizing thermal processes for nutrient retention. In addition, Ohlsson

(1980a, 1980b) found optimal sterilization temperatures for sensory quality factors in both cylindrical and flat containers.

OPTIMIZATION IN FOOD DEHYDRATION

All the case study examples presented thus far have dealt with optimization of thermal processing. This was a deliberate choice, based on the belief that readers could gain greater understanding from illustrations involving repeated use of the same process model in a variety of ways to obtain product and process improvement. Once a reliable mathematical model is available for simulation of any food industry unit operation, many of the techniques and approaches described in the preceding examples may be applicable for process optimization on that unit operation. Since a mathematical model for food dehydration has also been described in Chapter 5, this final section on process optimization will describe an application to food dehydration.

This example is taken from Mishkin et al. (1982), who used a mathematical model of the drying process based on Fick's law, which is similar to the process model for drying beet slices described in Chapter 5. The product to be dried was a model system composed of water, cellulose, and ascorbic acid (vitamin C) in the configuration of a thin, flat slab to provide rapid heat and mass transfer. This configuration permitted the assumption that temperature and moisture content could be treated as lumped parameters (i.e., parameters uniformly distributed at any point in time). The process consisted of air drying in a tray dryer, as described by Villota and Karel (1980a, 1980b).

The objective in this case study example was to solve a dynamic optimization problem in which the decision variable would be the air temperature as a function of time. The solution to the problem would be an optimum air temperature profile that would result in maximum ascorbic acid retention while achieving the desired final moisture content, given a specific' drying time. Formulation of this problem is very much the same as in dynamic optimization of thermal processing, described earlier.

The basic parameters needed for specifying the product and process model are shown in figure 6.12, where T_{db} and T_{wb} are the dry-bulb and wet-bulb temperatures, respectively, of the air stream; $T_{s,\bar{m}}$, and C are the temperature, moisture content, and ascorbic acid concentration in the product at any time; and L_1, L_2, and L_3, are the physical dimensions of the product slab. The rate constants used to specify the kinetic models for ascorbic acid degradation as a function of moisture content and temperature were determined by Villota and Karel (1980b) and are shown in figure 6.13. With such data available, the model could be used repeatedly through computer simulation to predict the final ascorbic acid retention for any given set of process conditions.

Two optimization techniques were used to carry out the search routine

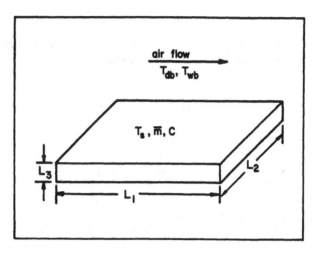

FIG. 6.12. Representation of the slab-shaped experimental product sample used as the basis for the food
dehydration process model in optimization studies. (From
Mishkin et al., 1982.)

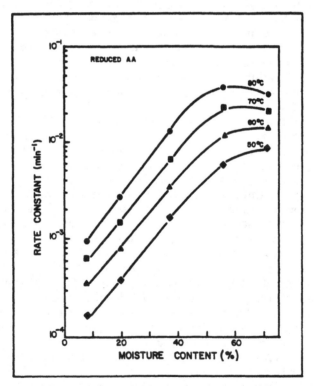

FIG. 6.13. Reduced ascorbic acid degradation as a
function of moisture content and temperature. (From Villota and Karel, 1980b.)

for finding the optimum temperature profile. The first method was based on the Pontryagin maximum principle as used by Saguy and Karel (1979) for the dynamic optimization of thermal processing, described earlier in this chapter. The second method was based on the complex method of Box (1965), which leads to a discrete-step optimum temperature profile.

Before using either technique, appropriate constraints on time, temperature, and slab thickness had to be established. This was accomplished by executing the model repeatedly over a range of different but constant dry-bulb air temperatures for different slab thicknesses. Figure 6.14 shows moisture and ascorbic acid retention and temperature of the drying slab for a simulated process at 75°C. These results for the simulated process are typical of the data acquired for real food systems such as sweet potato slabs. Figure 6.15 shows the results of the drying simulation for a series of slab thicknesses and constant air temperatures as indicated. It specifies the time required to dry from the initial moisture content of 2.5 g per gram of solids to the desired final moisture content of 0.05 g per gram of solids and the final retention of ascorbic acid for a particular combination of drying temperature and slab thickness. This single chart yields a considerable amount of information, serving as a useful precursor to optimization. From figure 6.15 it is evident that increasing the slab thickness demands longer drying times to achieve the necessary dryness and that this results in greater ascorbic acid loss. It is obvious that increasing the air temperature results in shorter processing time, but also at the expense of greater nutrient loss. These general results are intuitively obvious, but they serve as examples for situations in which similar insights are not readily attained by intuition. In addition, the quantitative results are not attainable by intuition.

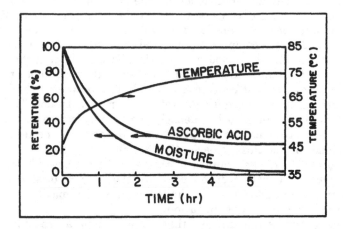

FIG. 6.14.   Drying simulation for a test slab of 0.6-cm thickness at constant dry-bulb temperature T_{db} of 75°C and wet-bulb temperature T_{wb} of 45°C. (From Mishkin et al., 1982.)

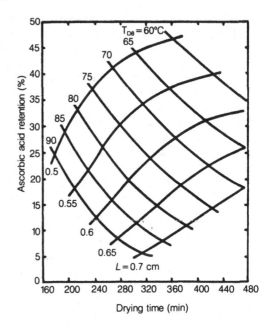

FIG. 6.15. Calculated ascor- bic acid retention and required drying times for different values of dry-bulb air temperature and slab thickness. The required final moisture content was 0.05 g per gram of solids in all cases. (From Mishkin et al., 1982.)

The region for decision search must be constrained to remain within the range of reliability of the models and the practical operating ranges of the equipment. The state equations described for the dehydration process were reliable within the air temperature range of 60–90°C; therefore, this was taken as the feasible search region for the optimal profile during drying.

Figures 6.16 and 6.17 contrast the optimal air-temperature profiles with resultant ascorbic acid retention during drying for 330 and 360 min using the discrete decision complex method and the continuous maximum principle. The discretized method may be visualized as finding the best configuration of five connected straight lines, while the continuous maximum principle finds the best continuous flexible line between t_0 and t_f. Since digital processing is inherently discrete, the term "continuous" in this context refers to a high-resolution, densely populated, discrete approximation to the true continuous analog. Given these differences, the results of the two methods are as close to each other as can be expected.

The resolution of the complex method may be increased by increasing the number of discrete decision points at the expense of substantial increases in computation time. This improved resolution may be unnecessary in most food engineering applications.

Common sense would suggest that the air temperature should be low during the early stages of drying in view of the high temperature sensitivity of ascorbic acid at elevated moisture contents, as shown in figure 6.13. As the moisture content is reduced, the nutrient is rendered less sensitive to

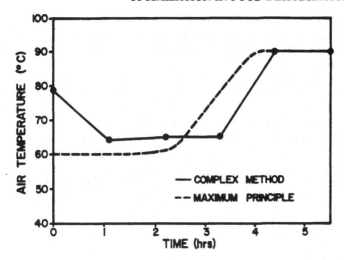

FIG. 6.16. Optimal profiles of the decision variable for maximizing ascorbic acid retention in drying, using two optimization methods. Drying period was 330 min, final moisture content 0.05 g per gram of solids, wet-bulb air temperature 45°C, and slab thickness 0.6 cm. Ascorbic acid retention by complex method was 27.8 percent, and by maximum principle 28.0 percent. (From Mishkin et al., 1982.)

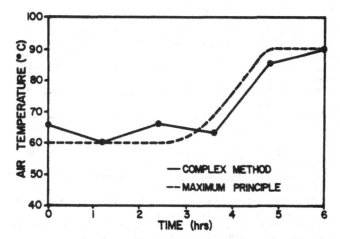

FIG. 6.17. Optimal profiles of the decision variable for maximizing ascorbic acid retention in drying, using two optimization methods. Drying period was 360 min, final moisture content 0.05 g per gram of solids, wet-bulb air temperature 45°C, and slab thickness 0.6 cm. Ascorbic acid retention by the complex method was 29.7 percent and by the maximum principle 30.7 percent. (From Mishkin et al., 1982.)

degradation and the air temperature may be increased. This is precisely the scheme depicted by the optimal profile in figures 6.16 and 6.17. It is coincidental that the usual moisture-temperature regime during hot air drying is naturally protective against ascorbic acid destruction, as can be visualized by referring again to figure 6.14. During the early stages of the falling-rate period, the rate of drying is sufficiently high to keep the drying material temperature low, which counteracts in part the sensitivity of ascorbic acid to degradation at high moisture contents. As the drying progresses, the temperature rises but the nutrient becomes more stable because of the falling moisture content. In view of this naturally stable regime for air drying of ascorbic acid–containing systems, use of the prescribed optimization technique improved retention of ascorbic acid by 5–10 percent over the comparable result obtained for the constant-temperature processes represented in figure 6.15.

REFERENCES

BOX, M. J.   1965.   A new method of constrained optimization and a comparison with other methods. Computer Journal 8(1): 42.

BROOK, R. C., and BAKKER-ARKEMA, F. W.   1978.   Dynamic programming for process optimization. I. An algorithm for design of multi-stage grain dryers. Journal of Food Processing Engineering 2: 199.

EVANS, L. B.   1982.   Optimization theory and its application in food processing. Food Technology 36(7): 88–96.

EVERSON, G. J., CHAN, J., and SHERMAN, L.   1964.   Aseptic canning of foods. Food Technology 18.

FELLICIOTTI, E., and ESSELEN, W. B.   1957.   Thermal destruction rates of thiamine in pureed meats and vegetables. Food Technology 11(2): 77–84.

HILDEBRAND, P.   1980.   An approach to solve the optimal temperature control problem for sterilization of conduction-heating foods. Journal of Food Processing Engineering 3: 123.

JOSLYN, M. A., and HEID, J. L.   1963.   Food Processing Operations, Vol. 2. AVI Publishing Co., Westport, CT.

KAREL, M., SAGUY, I., and MISHKIN, M.   1984.   Optimization methods and available software in food engineering. Part I. CRC Critical Reviews in Food Science and Nutrition 20(4).

LUND, D. B.   1977.   Design of thermal processes for maximizing nutrient retention. Food Technology 31(2): 71.

LUND, D. B.   1982.   Applications of optimization in heat processing. Food Technology 36(7): 97–100.

MARTENS, T.   1980.   Mathematical model of heat processing in flat containers. Ph.D. thesis, Catholic University, Louvain, Belgium.

MISHKIN, M., KAREL, M., and SAGUY, I.   1982.   Applications of optimization in food dehydration. Food Technology 36(7): 101–109.

NORBACK, J. P.   1980.   Techniques for optimization of food processes. Food Technology 34(2): 86.

OHLSSON, T.   1980a.   Optimal sterilization temperatures for flat containers. Journal of Food Science 45: 848.

OHLSSON, T.   1980b.   Optimal sterilization temperatures for sensory quality in cylindrical containers. Journal of Food Science 45: 1517.

PONTRYAGIN, L. S., BOLTYANSKII, V. G., GAMKRELIDZE, R. V., and MISH-CHENKO, E. F.   1962.   The Mathematical Theory of Optimal Processes. (In Russian; transl. by K.N. Tritogoff.) Wiley-InterScience, New York.
SAGUY, I.   1983.   Computer-Aided Techniques in Food Technology. Food Science Series, Vol. 8. Marcel Dekker, New York.
SAGUY, I., and KAREL, M.   1979.   Optimal retort temperature in optimizing thiamine retention in conduction-type heating of canned foods. Journal of Food Science 44(5): 1485.
STUMBO, C. R.   1965.   Thermobacteriology in Food Processing. Academic Press, New York.
TEIXEIRA, A. A.   1971.   Thermal process optimization through computer simulation of variable boundary control and container geometry. Ph.D. dissertation, University of Massachusetts, Amherst.
TEIXEIRA, A. A., DIXON, J. R., ZAHRADNIK, J. W., and ZINSMEISTER, G. E. 1969.   Computer optimization of nutrient retention in thermal processing of conduction-heated foods. Food Technology 23(6): 137.
TEIXEIRA, A. A., ZINSMEISTER, G. E., and ZAHRADNIK, J. W.   1975.   Computer simulation of variable retort control and container geometry as a possible means of improving thiamine retention in thermally processed foods. Journal of Food Science 40: 656.
THIJSSEN, H. A. C., KERKHOF, P. J. A. M., and LIEFKENS, A. A. A.   1978.   Short-cut method for the calculation of sterilization conditions yielding optimum quality retention for conduction-type heating of packaged foods. Journal of Food Science 43: 1096.
VILLOTA, R., and KAREL, M.   1980a.   Prediction of ascorbic acid retention during drying. I. Moisture and temperature distribution in a model system. Journal of Food Processing and Preservation 4: 111.
VILLOTA, R., and KAREL, M.   1980b.   Prediction of ascorbic acid retention during drying. II. Simulation of retention in a model system. Journal of Food Processing and Preservation 4: 141.

INDEX

Page numbers in *italics* refer to figures; page numbers in **boldface** refer to tables.

Printed in the United States
by Baker & Taylor Publisher Services